CHANGING CHANNELS

Issues and Realities in Television News

Jerry Jacobs
California State University, Northridge

Mayfield Publishing Company
Mountain View, California
London • Toronto

Library of Congress Cataloging-in-Publication Data

Jacobs, Jerry (Jerry M.)
    Changing channels : issues and realities in television news /
Jerry Jacobs.
        p.   cm.
    ISBN 0-87484-946-2
    1. Television broadcasting of news—United States—History—20th
century. 2. Television broadcasting—Social aspects—United States.
3. Cable television—Social aspects—United States. 4. Television
broadcasting policy—United States—Evaluation. 5. Broadcast
journalism—United States—Objectivity.   I. Title.
PN4749.J25   1990
070.1'95—dc20
                                                            89-29964
                                                            CIP

Manufactured in the United States of America
10 9 8 7 6 5 4 3 2 1

Mayfield Publishing Company
1240 Villa Street
Mountain View, California 94041

Sponsoring editor, C. Lansing Hays; managing editor, Linda Toy; production editor, Carol Zafiropoulos; manuscript editor, Carol Dondrea; text and
cover designer, Jean Mailander. The text was set in 10/12.5 Palatino and
printed on 50# Glatfelter Spring Forge by Thomson-Shore, Inc.

Contents

Preface

The purpose of *Changing Channels* is to provide the college journalism student with an up-to-date, on-the-scene, realistic, issue-oriented portrait of broadcast news industry policies and practices. The book's subjects are the people, professional standards, ethics, commercial demands, and technologies that shape today's and tomorrow's TV newscasts.

Today's broadcast journalism students are tomorrow's industry leaders and will help determine—ethically, practically, and professionally—the directions in which the industry moves. It is important for them to understand both the history and the present state of broadcast news in order to make their contribution to tomorrow. Especially, they need to know that despite growing pains, despite happy talk and blatant commercialism, broadcasting has a history of meaningful journalism that is genuinely concerned with societal needs.

Changing Channels has been inspired in part by the Chicken Littles who cry that the industry's sky is falling. The persistent refrain from former network news executives and others is that network news, with its declining ratings and diminishing number of viewers, is dying—victim of increasingly powerful and ratings-crazy local stations whose news is dominated by sex, sleaze, and commercialism.

That there has been remarkable change is undeniable, but it is more a matter of the weather changing than the sky falling. Change, basic in life, is particularly characteristic of broadcast journalism because broadcast journalism is both electronic and concerned with news. Because the changes over recent years have been so dramatic and so visible, I felt the need to go directly to the source—to industry practitioners and principals—for a realistic appraisal of the current state of broadcast journalism. The object was to record their answers in a format—this book—that college students could use to decide for themselves whether the sky is clear, darkening, or about to collapse.

The topics chosen for *Changing Channels* are broad: high technology, ratings, sweeps, network relations, local, national, and international content, coverage parameters, the influence of consultants, news programming, anchors' runaway salaries and dominance, the quality of personnel, improving the product, a look ahead for local and network TV news, preparing for the job market, and the immediate expectations and apprehensions of employers regarding college student training and entry-level skills. I explore these issues in order to describe the television news professional's real world. The goal is to provide a text that can supplement, with a realistic description of present practices, more skills-oriented books on news writing, reporting, shooting, and editing techniques. It is also intended to arouse student interest for further discussion, debate, and exploration.

Changing Channels incorporates findings from a wide cross-section of large and small stations throughout the United States. Trends and practices were studied in both the more representative but less obvious markets and in the "golden ghetto" markets of the larger metropolitan areas.

I interviewed overseas TV journalists and observed their operations to determine whether Americans could learn something new by comparing their own style of broadcast journalism with the styles of other cultures, economies, and political systems. I was also interested in ascertaining to what extent our good and bad habits were being exported to foreign countries.

Altogether, the odyssey covered 23 cities in six countries and included face-to-face interviews with more than 50 news executives. Where possible, printed sources are documented, but the main sources were interviews with news directors, executive producers, managing editors, producers, anchors, reporters, writers, assignment editors, technicians, station managers, network executives, news organizations, and federal officials.

Each chapter ends with a summary and a projection concerning the future evolution and development of the particular aspect of broadcast journalism discussed.

As in any profession, a certain amount of jargon is unavoidable, and broadcast news students will benefit by becoming familiar with it. Specialized terms are shown in boldface type the first time they occur, and, when feasible, they are defined in the body of the text. In addition, there is a glossary at the end of the book.

Acknowledgments

I am indebted to a number of organizations for financial assistance. The Gannett Foundation was the primary sponsor, but supplemental funding came from California State University, Northridge, the CSUN Foundation, and the H. V. Kaltenborn Foundation.

A number of my colleagues contributed their opinions of this manuscript in early drafts. I wish to thank Jim Bernstein, Indiana University; Donald G. Godfrey, Arizona State University; Dan Gingold, University of Southern California; Tom Grimes, University of Wisconsin, Madison; John Hewitt, California State University, San Francisco; Benjamin Holman, University of Maryland; George A. Mastroianni, California State University, Fullerton; and Roger L. Walters, California State University, Sacramento.

I would like to thank the good and patient people at Mayfield Publishing: Lansing Hays, sponsoring editor; Linda Toy, managing editor; Carol Zafiropoulos, production editor. Thanks also to Carol Dondrea who copyedited the manuscript.

To the person who held my hand
over the rockiest terrain,
my wife,
Nancy Baker Jacobs

INTRODUCTION

Our century, the 20th, with all its electronic bells and whistles, is gradually fading to black. In video terms, that means its images, along with the people who created them, will soon start disappearing from the screen. And, as in television, they will **dissolve** to other pictures and different sounds to be determined by fresh minds and faces.

This work was designed for them, the developing broadcast journalists for whom this particular camera scans, reports, and projects. The view includes both sides of the camera, the on-the-air product as well as the off-the-air process of local television news. In reporting these scenes, my attempt is to emulate the camera's uncompromising delivery of stark truth. It is an attempt to bring into focus the "gee-whiz" wonders as well as the blemishes. The subject is a product that

1

has become virtually a way of life for millions of people—the news-viewing junkies—and a big business worth billions of dollars. By the last page, readers should have a clearer sense of broadcast journalism today. As in "A Christmas Carol," there will also be frequent glimpses of broadcast journalism past, when appropriate, and broadcast journalism future. The analogy is not a frivolous one. Ebenezer Scrooge wasted little time changing his world after receiving his gift of vision.

However, before contemplating changes in TV news it is important to be armed with appropriate information. That is one of the goals of *Changing Channels*. The facts, to a certain extent, do support the position of the Chicken Littles. Although not all of broadcast journalism is a wasteland, as they contend, certain aspects of TV news are in trouble and need immediate attention.

Now, in the spirit of substantive TV news teases, a sampling of the results:

- Local TV news operations are making more money than ever but those who control budgets are doling out less to make it happen.
- High tech is sweeping the United States more rapidly than other countries, including even Japan.
- The biggest channel change is in local TV news, which is now capable of covering news anywhere, regardless of what the network does.
- Network TV news is abdicating to local operations, even to the point of depending on the affiliates for coverage of breaking major stories.
- There are so many new sources and opportunities to cover out-of-town news material that so-called local-local news content is suffering.
- There is major concern about the quality and motivation of broadcast journalism newcomers.
- The art of TV news writing—the heartbeat of any TV news product—is in critical condition.
- Despite the prosperity of local news operations, the late afternoon news programming block will shrink.
- The number of local newscasts will grow, and these will be spread out over the entire day, with the early morning and midday periods the immediate targets.
- The 11 o'clock news is ailing and will die in some markets.

- Sports segments on local newscasts will have to undergo a dramatic change or they will also be in the obit column.

ℓ
- While many stations will do anything for ratings, there is plenty of old-fashioned journalistic integrity still out there.

- *Most* local news directors recognize a need for network news, but visualize customizing it for themselves.

- It's always the other guy who is doing the sex and sleaze.

- The old principle that broadcast journalists, not the public, determine what should go into newscasts is dead but not quite buried.

- Somewhat galvanized news consultants, parasitic descendants of the 1970s, continue to multiply and influence news operations in the 1980s.

- Everybody, except the anchors, is concerned about the dominance of the anchors and their runaway salaries.

X
- Most foreign broadcast journalists have a deep respect for depth in reporting.

X
- TV and radio journalists cooperate successfully overseas in covering stories as well as in creating unique programs.

x
- Certain news operations in other countries consolidate jobs, resulting in better on-the-air product.

- Many foreign operations, particularly newer ones, imitate our good and bad news habits.

- Private enterprise is wresting control of government-owned and government-run TV operations overseas.

As the student undoubtedly knows, broadcast journalists don't have the security expectations of, say, doctors, lawyers, or funeral directors. Even test pilots, frogmen, movie stuntmen, and demolitions experts last longer in their jobs. The cold reality is that news directors average *two years* at any one station. In other words, the executives quoted in the pages that follow are more than likely to have different titles, call letters, and addresses today than they did when I interviewed them. In addition, in some instances, equipment, plans, and personnel may have shifted direction from yesterday to today. The choice I faced then was committing to the way it was at press time or continually revisiting the interview subjects. Since it was unlikely any philanthropic foundation would support the latter direction, I opted for the former.

And *now* the details. . . .

1

---◼---

TOYS *WERE* US?

The Impact of Technology
on TV News

Broadcasting live from anywhere any time is having tremendous impact on today's and tomorrow's television journalism. But that doesn't mean every broadcaster has every tech toy. Nor is every newsroom playing with these techniques maturely.

It was a news producer's nightmare. Our TV news specials unit at NBC News in New York was set to hit the air with the first of a series of programs on the Big Four Summit in Paris. Just prior to air time, the Khruschev–Eisenhower–Macmillan–de Gaulle historic meeting went down in flames, just like the problem that caused the failure—the American U-2 reconnaissance spy plane caught in the cross hairs over the Soviet Union.

And thus the concept **instant news specials** was born. Rather than cancel out the prime time special, our unit improvised. We quickly called up radio lines to the European capitals, energized our audio/video line to Washington, D.C., and put our fastest editors to work reediting all those now-outdated film stories we had labored over for days.

The result was mostly old-time radio on TV, with anchors in New York talking to reporters overseas on radio circuits. While we were on the air, editors and writers rushed through the laborious, time-consuming process of searching, reediting, and rewriting the film stories. These were then quickly interspersed with the interviews. Despite the radio characteristics, kudos poured in and we news commandos were particularly proud of ourselves. We had shown off to the world, and particularly the competition, both our ability to react and our technological bag of tricks. The show went on.

That was in May 1960, the era of ordering up a video line (only within the United States, of course) and hoping for the best. Some lines—between New York and Washington, for example—were preset and could be activated instantly. Others could take anywhere from a few minutes to weeks, depending on where you wanted to go. Images were recorded on film, which meant it had to be shot, messengered to the lab, processed at a rate of approximately 20 minutes a roll, screened, and edited. Before the film stories could be projected there was still another step: Writers had to prepare narration for the anchors to read with the film. The shooting aspect was usually a given in that we hired the best. But actually getting it on the air depended on everything from the speed of the bike messenger and the reliability of the flight captain to whom we entrusted the can of film to the weather conditions for flying, whether connecting flights were caught, whether chemicals used to develop film in the lab were accurately measured, whether editors and their equipment were scratch-proof, the IQ of the writers, and the temperament of the projectionists and their equipment.

As for the frills—graphics, animation, and captions—they too could take anywhere from a few minutes to a few hours to days. They were done by a contracted company in a different building. Producers had a fighting chance with the captions, or **lower-third supers**, as they were known—information such as names, titles, dates, and addresses appearing on the bottom part of the TV screen. However, actual artwork, such as portraits of newsmakers, show titles, maps, charts, and animation that really was animated, had to be conceived and ordered days in advance.

There were also special typewriters for **TelePrompTers,** devices that project the script in front of the camera lens for the newscasters to read, but they were so slow that using the machines for a breaking instant news special was out of the question.

And there were other horror stories. The vital **show rundown,** or **lineup,** which is the detailed show outline, including story elements and running times, was handwritten or typewritten. The all-important timing of each element and of the overall show we did in our heads or, many of us, on our fingers. As air time approached and show elements changed or were dropped, the rundown inevitably changed, which meant doing it all over again and several trips back to the Xerox.

For just about every one of the last 30 or so years since the summit that did not take place in Paris, Santa has been good to the TV news business. In the 1960s, he gave us color film, then videotape. At about the same time, he took away our manual typewriters, replacing them with electric typewriters, and finally, computers. **Graphics** went from laboriously hand-drawn visuals to rear screen projection to today's magical **paint boxes,** which give news operations the capability of Hollywood animation. It's the kind of razzle-dazzle that seduces viewers at the beginning of the news show or keeps them tuned during it. For example, it can give the audience a feeling of flying over the White House or Congress when, in fact, the "buildings" are computerized models. Getting the viewer's attention is the name of the game, and the method is in-and-out zooming movement, machine gun delivery of information, and a hint of 3-D.

ENG, SNG TO THE RESCUE

The technological development that is having the greatest, near revolutionary, impact on the content and direction of TV news is satellite capability. The first major step in that direction was the advent of **ENG, electronic news gathering,** in the 1970s. At that time, lightweight, compact cameras and videotape decks finally took those heavy, bulky film cameras off the shoulders of camera operators. With ENG came the ultimate escape from cumbersome, expensive cables on the ground to the airwaves and microwave transmission.

In the 1980s, the next step happened inevitably: **SNG, satellite news gathering.** Local stations acquired vehicles from which to shoot pictures to satellites, which bounce them back down again to a ground receiver. These **SNV,** or **satellite news vehicles,** have their own mysterious vocabulary. They are often referred to as **Ku-band trucks. Ku-band** simply refers to the section of the electromagnetic spectrum assigned to satellite use. The signal shot from a transmitter, or Ku-band truck in this case, to a satellite is called an **uplink**. A **downlink** is a signal from the satellite to the ground receiver.

Most SNVs are virtually self-contained. They carry editing equipment, cellular telephones, **microwave** capabilities, production **mixers**, and computers, which give them writing, communication, and research capabilities.

To **producers**, the people responsible for putting on the local and network news shows, all this sophisticated equipment means *instant*

- capability to originate a broadcast from any breaking story virtually anywhere in the world at any time of day;
- painless computerized writing and editing;
- modem-generated availability of research materials in the field;
- computerized, no-headache show outlines;
- computerized, almost foolproof TelePrompTer services;
- graphics, with or without animation, covering any eventuality;
- and a Big Brother-is-watching-you computerized assignment desk operation that means quick as lightning access to all staff members, all the time, everywhere.

THE OVERVIEW

The wizardry also means that the networks no longer have a monopoly on covering major events. An event such as the presidential inauguration of 1989 drew some 40 Ku-band trucks to Washington from all over the country. Some of the vehicles originated exclusive, custom-made coverage for stations affiliated with the networks. Others, such as Group W's Newsfeed, a service provided by the corporation made up of Westinghouse Broadcasting and Cable Inc. stations, rented out its facilities to crews from 15 stations and fed the event to 95 others. Groups of television stations could also subscribe to a pool produced by C-SPAN, the 24-hour cable system airing public affairs programming. Eighteen organizations took advantage of that bargain, three days of inaugural events for a mere $700.

Originating complete network *and* local newscasts from the national political convention in Atlanta or New Orleans, the Sugar Bowl in Miami or the Olympic games in Korea has become routine. The three networks, Cable News Network, Ted Turner's all-news network, and even a handful of local stations originated from China during the dramatic student turmoil and from Cuba during Soviet leader Mikhail Gorbachev's visit in 1989. After the death of Japan's Emperor Hirohito in 1989, CBS shelled out $2 million in order to originate funeral coverage as well as its evening newscast, "CBS This Morning," "Sunday Morning," and "48 Hours" from Japan. Although the other networks weren't quite as ambitious, viewers could nevertheless choose to watch the events from among several Tokyo locations: ABC's Peter Jennings anchoring from above a train station, NBC's Tom Brokaw in a Japanese garden, and the CNN team of Mary Alice Williams and Bernard Shaw from the Okura Hotel.

Again, stations and less opulent networks could take advantage of the satellite companies' services for customized TV news coverage. Conus Communications, the satellite-energized mini-network operating out of Minneapolis, fed stations in Boston, Richmond, Virginia, and Long Island, New York, as well as the "MacNeil, Lehrer NewsHour," cable's Financial News Network, the Canadian Broadcasting Corporation, the Spanish language Univision Network, and the Independent News Network.

In other words, any event anywhere in the world can now be covered in some manner by any size TV station.

It's exciting, this new gold mine of news coverage. But to many national news leaders it is also scary.

Robert Mulholland started as a news writer, and as he progressed— to producer, news director, and, ultimately, president of the NBC network—so did high tech. At the time of this survey, Mulholland was president of the Television Information Office, network TV's booster club located in New York. He admits broadcast journalism had growing pains during its electronic metamorphosis. Yet, he says, "We're at the point now where we're going to start to control the toys and not let the toys control us." Mulholland, who saw the change from black and white to color, from film to videotape, and from **VTR** to live, is optimistic about the use of the satellite news gathering. He feels stations use satellites to improve regional coverage. "Because you have a satellite truck doesn't mean you're covering news in Geneva."

A NATIONAL PERSPECTIVE

The National Association of Broadcasters in Washington is an umbrella association representing stations, management, and owners. NAB has been active in a myriad of movements involving TV news, such as fighting for First Amendment rights and against the Fairness

Doctrine bill. Among its many research subjects is the field of technology, with emphasis on the trend of local news departments toward using satellite news vehicles.

According to NAB, 122 TV stations are using satellite news-gathering vehicles, an increase of 72 percent in less than two years. The greatest expansion has been in the small- to medium-sized markets. The research department of NAB also reports 98 percent of stations using SNVs belong to one or more private regional or network satellite news-gathering cooperatives or services.

About half of the SNV users, according to NAB, are affiliated with Conus. The others are divided among the National Broadcasting Company, the Columbia Broadcasting System, the American Broadcasting Company, the Cable News Network, Group W, and regionalized groups of stations.

NAB vice president Walter Wurfel sees good and bad in the satellite craze for local stations, depending on how the vehicle is used. He maintains that using it to produce features "around your own state can be an enhancement of local news coverage." However, "to the extent that local stations send one of their correspondents to Lebanon, it's questionable."

Ernie Schultz, president of the Radio-Television News Directors Association in Washington, is quick to admit his is not the organization to consult for criticism of local TV news. RTNDA is made up primarily of news executives throughout the United States. Characteristically, Schultz is enthusiastic about the bird (a slang expression for satellite) and everything it can do for news. Schultz, a 25-year TV news veteran, sees local news finding its place covering national and international news "with a strictly local angle." Asked whether revolutions in foreign countries, which some local TV stations cover, would fall into that category, he responds, "Absolutely. Once it all shakes down, if there's a local angle, they'll find it."

However, in many respects the satellite has given local stations a blank passport, transporting them farther and farther from City Hall. That problem will be taken up in greater detail in subsequent chapters.

AN ABUNDANCE OF PRODUCT

The new technology has created a maze of news feed services and hardware products that could warm the cockles of the coldest of producers' hearts. Available to the stations are feeds from the mini-networks created by Conus Communications, Cable News Network, Financial News Network, C-SPAN, the Weather Channel, regional networks, consortiums that cross affiliate lines into super-networks, ad hoc alliances thrown together for the purposes of any one breaking story, and, significantly, the parent networks themselves.

So far there are no traffic jams in outer space, but there is gridlock on earth. So fast, so frequent, and so full are these incoming feeds that the stations are having trouble supplying the personnel to monitor and prepare them for on-air use. Available practically around the clock, the resulting material can easily fill the newscast of any size station without the need for any locally produced news.

A small station like WPBN-TV in Traverse City, Michigan, the 139th largest market in America, has more satellite news feeds on a daily basis than it has news staffers. There are only 18 employees, but a minimum of 20 feeds. Six come from the NBC network, nine from CNN, and five from Conus.

The networks no longer jealously guard what they used to consider as theirs exclusively, the right to cover national and international stories. Many news veterans see these concessions given to local stations in news coverage as a complete abdication on the part of the networks. For one thing, stations don't even have to pay the entire cost of the half-million-dollar Ku-band trucks. The networks, anxious to pacify increasingly independent affiliates, go in fifty-fifty with them.

NETWORK SYNDICATION

One of the most dramatic changes in the relationship between network and local news is in the syndication area. **Syndication**, composed of news, sports, and feature stories, is the service provided by the networks to the stations for use in their local newscasts. Not too long ago, this service was made up of the network news department dregs, the stories that—for a variety of news and professional reasons—didn't make the big shows.

Today, the mood and the product are vastly different. Paul Dolan is a key executive in NewsOne, the ABC News Network Syndication Service. His goal: "To provide [ABC-affiliated] stations with the best national and international news pictures possible." To accomplish that, Dolan and his colleagues have created news cooperatives "so that stations are really helping one another." The result is an all-day service with live **feeds** when necessary and six exclusively regional feeds. This translates into some 180 pieces of video daily for the **affiliated** stations.

No longer do networks send out material on a take-it-or-leave-it basis. At ABC headquarters in New York City, a morning conference call rings in all bureaus from around the world and the United States. The morning is then spent polling the various TV stations for what's happening and what is needed around the country. Another conference in the early afternoon brings together all ABC regional managers and staff. These stations can receive video items on the regional feed, as well as show lineups, or outlines, and actual scripts, which are available from the network's main computer.

Dolan insists that, at ABC, the SNVs strengthen the network competitively, despite the station's control over the mobile units. "We initially did want to have the right to say, 'You've got to move that truck over here.' That is not acceptable to the stations, which felt that they had to have complete control." In other words, ABC—along with NBC and CBS—is taking a far more realistic approach to the burgeoning affiliates.

CBS and NBC have similar services, with the latter feeding from 11:30 A.M. to 2 A.M. the next morning.

Jo Moring, vice president of NBC's Syndication Service for Affiliates, confirms that providing an uplink/downlink pathway through the sky encourages stations to dispatch crews worldwide. "If you're in Houston and want to go to Austin to cover the state legislature, you can do that. And if you're in Houston and want to cover New York, you can do that, too."

Like her counterparts at the other networks, Moring's daily news menu reflects the emphasis on national coverage, including hard breaking news, interviews, sports clips, national satellite weather maps, and a variety of graphics.

For a small station with a meager budget, network services can be a godsend, providing the affiliate's management takes advantage of the windfall to concentrate money and personnel on local news coverage. However, many stations are using the material instead to produce broader newscasts, particularly if they have the use of Ku-band trucks.

THE SURVEY STATIONS

In terms of "area of dominant influence," broadcast jargon for the size of the station's market area, cities visited for *Changing Channels* ranged from heavily populated San Francisco, in the top five, to Helena, Montana, in the bottom five. The big stations, such as NBC's KRON-TV and ABC's KGO-TV in San Francisco, naturally have it all: the newest and best in high tech, a generous choice of feeds, and a staff bigger than the entire NBC network news team of the late 1950s.

KRON-TV **news director** Herbert Dudnick has a staff of 140 people, giving the station 18 shooting crews and bureaus in the East Bay, South Bay, Sacramento, and Washington, D.C. In addition to the fountain of news feeds from NBC, Dudnick's newsroom also is part of the Conus hookup, which means it has an SNV truck and shares the output of its 80 or so stations.

Dudnick, a veteran broadcast journalist, whose NBC Network News credits date back to his writing days for the Huntley-Brinkley News, is convinced that technology no longer overwhelms TV news. At one time "KRON did a lot of live shots" just because it had the capability. But now, insists Dudnick, "at least in this operation, we cut way back in the amount of live coverage. We only go live when absolutely necessary." An example of "absolutely necessary" live coverage was the airlifting of troops from nearby Fort Ord, 85 air miles south of San Francisco, to Honduras.

Unfortunately, KRON-TV had major difficulties going live when one of the biggest stories of the year hit home—San Francisco itself—on October 17, 1989. While key competitors went on the air live with coverage of the 7.1 killer earthquake, the NBC station remained dark for the first 30 minutes because it was knocked off the air. To further complicate matters, confusion at network headquarters in New York delayed NBC nationwide coverage another half hour. The other networks in the meantime aired the dramatic, locally produced coverage of their San Francisco affiliates, further embarrassing NBC and KRON.

ABC and its owned and operated station, KGO-TV, was in the ideal position of having the best in technical equipment on hand and oper-ational for the World Series coverage at Candlestick Park. When the temblor occurred during the pre-game program, ABC and KGO-TV were able to segue deftly into quake coverage, beating both NBC and CBS.

Ironically, KGO-TV had had a staff of 148, but that had been whittled down in the months prior to the quake by the sharp pencils of the new owners, Capital Cities-ABC Inc. Harry Fuller, who has been in the news business since graduating from Stanford in 1969, has so many incoming feeds that he doesn't have the people to juggle them properly. As a member of CNN, KGO has a constant flow of material; it also has the ABC network feeds, as well as a service among the six affiliates owned and operated by ABC.

Of the news director's 16 microwave-equipped ENG units, eight are capable of going live and soon will be complemented by a Ku-band truck and a fully computerized newsroom. For graphics, KGO has the top of the line: two Quantel "paint boxes" capable of producing any combination of graphics in moments.

Fuller, like so many news executives, suffers from a nasty, but not necessarily fatal, side effect of the high tech bug. Management says, "Hey, we've spent a lot of money on your electronic marvels. So we want to see a live shot on the news all the time. Promote it and do it whether there's a story there or not. And if it's a satellite shot, all the better." This isn't a new story—other managers at other times told news directors to shoot flower shows to show off the newly acquired ability to shoot in color film. But mature news directors like Fuller do well in not letting themselves be pressured into going live unnecessarily. Fuller counters by telling the brass that "some of the worst things on television are live" and that live coverage doesn't allow you the opportunity to correct any mistakes. With luck, the glitz merchants will grow out of it, but who knows what other demands will come from future technological miracles.

Up the coast, in Portland, Oregon, KGW-TV's SNG truck has a full editing suite and two ENG trucks with live capability. In addition, two portable transmitters can activate any of the newsroom's ten camera units. Producers can also pick from all the national, international, and regional goodies from NBC Syndication. As one of the four TV stations owned by the prestigious King Broadcasting Company in the Northwest, KGW-TV can also tap into material from the other three cities (Boise, Seattle, and Spokane), including a daily 4 P.M. feed between KGW and KING-TV, the corporation's flagship station in Seattle.

Nepotism pays off. Thanks to the regionalized association, KGW was able to come up with a live shot of drought-stricken Kentucky receiving hay shipped there from Washington, and it had crews in Los Angeles and Sacramento contributing to its TV special on gangs. Managing editor Matt Shelley's people edit and originate from the affiliates, "so when a big story breaks we don't wait for the network. Local stations will cover it."

KGW news has a staff of 60.

In Seattle, the staff size of KGW's sister station reflects the larger market. KING-TV has 34 more people and the same feed capability as KGW. News director Donald Varyu realizes SNG has limited potential for him. SNVs, he says, are "good only as far as you can drive." Seattle is hemmed in by water to the west, has a sparse population in the Peninsula to the south, an international boundary to the north, and mountains to the east.

KING is among stations already computerized. The system, Basys, has been in place for a year and is already lashed up to TelePrompTer;

that is, it can already feed script materials from the computer directly into the TelePrompTer. The same computer system is expected to be rolling videotaped news segments directly into the **casts**, or newscasts, soon, saving several technical steps in that operation.

There are seeds of counterrevolution at KING, with the newsroom among the first to react against the surplus of material. Varyu dropped the CNN service and is proud of his decision. TV news success, he claims, will depend on how stations utilize technology to cover local news better—and local news is where KING's money goes.

OTHER BENEFITS FROM HIGH TECH

Technology can improve the quality of the product as well as its physical look. Producers and news directors, for example, now have a virtually painless opportunity to concentrate on the backbone and heartbeat of any news product: the writing. Copyreaders can get scripts faster and correct them more easily on the computer. Thanks to portable laptop computers, printers, **modems**, and FAX machines, reporters can send their copy from the field back to the studio by cellular phone and tap into any number of wire services and research banks.

James L. Boyer of KOMO-TV in Seattle uses such high tech to work with his reporters on script problems even several hundred miles away. The news director says the quality of writing in his shop has unquestionably improved.

Boyer's NewStar computer system will also soon control TelePrompTer, the playback of news stories, the stacking and playback of graphics and captions, and the rundown.

KOMO's staff of 90 persons has a jump on the competition with one of the first Ku-band trucks in the Seattle market, plus a microwave van with a 40-foot mast, and 10 news cars with microwave capability.

The ABC affiliate is also considering cutting back on services, in this case membership in the Conus network. Boyer would rather invest KOMO's energy and money in the mini-network of ABC affiliates that have satellite trucks. KOMO, well supplied by the ABC NewsOne feeds, already trades stories with KATU-TV, the ABC station in Portland.

KOMO, like all three network affiliates in Seattle, also has a helicopter, but, as is the case with many stations in the country, since the advent of ENG and SNG, it is used for other duties: transportation, traffic reports, a limited amount of aerial photography, and showing the station's flag with its colorfully displayed station call letters.

KSL-TV in Salt Lake City has everything but that new car smell. Owned by the Bonneville International Corporation, a Mormon Church enterprise, the CBS affiliate is a showcase. In addition to a new, modern facility and a Ku-band truck, KSL has 10 live units, a live truck, a helicopter, and one of the first computerized news operations in the country.

J. Spencer Kinard, vice president and news director, also has innovative equipment with which to originate live from any of his sedan-sized news vehicles. His engineers came up with new, lightweight microwave gear that can work off a tripod. The kit, which fits into the trunk of a car, gets whipped out at the news scene, creating a live capability from any location.

The computer system is NewStar, which KSL uses to streamline writing, rundowns, and assignments. Eventually it will feed Tele-PrompTer and news cassettes.

Kinard, who has held his title since 1972, reflects with little fondness on his earlier research and development venture into computerization. The computerization worked fine until the day his staff couldn't get the newscast out of the computer 20 minutes before the show. It was the only time Kinard ever had to delay a newscast—every producer's nightmare.

In the 16 years since the early experimentation KSL perfected its computer performance and installed the NewStar system. By the summer of 1988 the situation was dramatically improved. Salt Lake City was directly hooked up to Atlanta and the station anchored its newscasts from the Democratic National Convention. So smooth was the computer, satellite, and phone communication that scripts and rundowns originated in either location. As far as the viewer was concerned, the switching and on-air performance were as clean as if they were routine.

Financially, KSL-TV, with an annual **news budget** of $4 million, is doing well. But, like others, the station is pressured by the people who overseee the budget—**bean counters**, as they are kown in the trade. Kinard's staff of a hundred, for example, had to be cut by 20, but he still retains 15 reporters, the same number of photographers, nine producers, 10 editors, and nine anchors.

Kinard is also trimming his number of feeds, dropping Group W's Newsfeed and the Local Program Network, or LPN, which has the input of 40 affiliates and independent stations. He hasn't used them enough and couldn't justify the cost, but that still leaves him with CNN and CBS. However, KSL's satellite truck has made the station a lot of "friends who cross network affiliate lines." These informal arrangements with other stations provide KSL with still additional elements for the newscasts.

Kinard is among those who see the perils of high tech. The news executive feels that some broadcast journalists have "let the technology, this wonderful technology, dominate too many of the things we do for the wrong reason." These tools, he says, should be used "for bringing the information to people, not just to put the glitter on them."

CBS's affiliate in Boston, which is among the top 10 markets in the country, has, in addition to the customary hardware, the powerful New England News Exchange to draw upon. Jeff Rosser, news vice president of WNEV-TV before he was promoted to a general manager's job in Birmingham, Alabama, calls the Exchange unique. It includes New England newspapers and radio stations as well as TV stations. The mini-network, reminiscent of the better overseas operations, even boasts an ABC affiliate. The others are all CBS. Thanks to a microwave link, NENE is able to go live from just about any location within a six-state region. This, plus WNEV's satellite-connected KU-band truck, gives the station optimum mobility. The Boston station can also originate from any one of four regional daily newspapers that have microwave and videotape editing equipment. The staff of WNEV comprises 108 newspeople, including 27 reporters and anchors.

WTVJ in Miami may have had an identity problem—at one point the station was owned by NBC's General Electric and still carried CBS programming—but that hasn't prevented the station from keeping up with the tools of the trade. As a member of the Conus network, **assignment editor** Gary Mehalik has a satellite truck as well as three ENG trucks in Miami and one in Fort Lauderdale.

The station also has a variety of sources to bolster its 35 reporters and camera people. WTVJ has both CNN *and* Conus. The CNN newsfeed is piped in constantly and the Conus satellite network expands WTVJ news origination by thousands of miles. Add to these an array of feeds from the CBS network, which includes a Southeast Regional feed, and you should have the makings of happy newscast producers in the ideal position of having too many news elements from which to choose. However, the situation creates a problem for Mehalik because he feels obligated to cover stories anywhere in the world if there is a Miami angle.

WTVT news director Jim West is proud of his Tampa, Florida, ENG units, one of which had enough features—on-board editing, cellular phones, and a computer—to qualify as a mobile newsroom.

In St. Petersburg, Florida, ABC affiliate WTSP-TV has its pick of several network feeds as well as the Florida News Network. FNN is made up of three ABC stations and one CBS affiliate. That network by itself provides 200 to 300 news, feature, and sports stories a month. In addition, ABC's NewsOne feed gives the station excerpt rights on "World News Tonight" and a Southeast regional feed.

The news director of another ABC affiliate, KSTP-TV in Minneapolis, reports that he is practically overwhelmed by the number of feeds. Larry Price has around-the-clock news feeds from the network and Conus. Price, who hired a full-timer just to monitor the feeds, describes Ku-band as changing local news more than anything in recent years. "There's an unbelievable change in the amount of power available. You can get stories from anywhere." His favorite example of this was the crash of a Northwest Orient Airlines plane. Conus stations in Minneapolis, which is the airline's headquarters, Detroit, which was the scene of the crash, and Phoenix, the plane's destination, pooled reporter pieces, live shots and videotape.

Competing WCCO-TV, with its staff of 80, also has a wealth of news material. As a CBS affiliate, the station has access to the network's regional news feed. WCCO-TV is also a CNN affiliate, which gives it a comfortable bank of non-local visuals. In addition, WCCO is a member of the Local Programming Network, a service that it started. News director Reid Johnson, who has been with the station since 1972, describes the product as mostly evergreen, series reports on general subjects that can be used any time.

Ian Marquand's staff at KTVH-TV in Helena, Montana, is only a fraction the size of that at WCCO. The station makes do with one ENG unit, one **editing bay**, and one makeshift van, and the staff of four, including news director Marquand, do it all. They cover, shoot, write, edit, and anchor news, sports, and weather. KTVH has no photographer per se, so reporters shoot themselves for their own on-camera pieces, which are called **standuppers**. Marquand, for example, has specific places mapped out around the state capital where he knows he can set up the camera, get in position, turn on the camera, and race back into the picture.

It's not quite that simple on a field story, however. Caught alone on one such **breaker**, or developing story, former reporter-anchor Ashley Webster solved the problem of working alone by marking off his height against a telephone pole with chalk. In other situations, Marquand and Webster shoot each other's standuppers. It's also not unusual for them to ask the newsmaker to please check the camera's eyepiece for them before hitting the button.

For Marquand, the feeds he gets from NBC are a necessity. His newscasts include parts of the network's "Nightly News," the stock market report, weather graphics, some of the sports feed, and the headline stories of the day.

BACK TO EARTH?

KTVH isn't the only station that has trouble reaching for satellite transmission. The birds are becoming increasingly expensive, which opens up competition on the ground. **Fiber optics** cable is one possible alternative. It's a cheaper system, a fraction of the cost of satellite transmission, and capable of carrying billions of data bits across oceans or towns. The earthbound cables are also far easier and cheaper to repair than the outer space satellites. For the present, users see fiber optics only as a conveyance specializing in computer data, but it wasn't that long ago that news directors never would have envisioned originating their local sportscasts from the Super Bowl halfway across the country either.

AN EXPENSIVE GIFT FROM ABROAD

It's difficult for TV old-timers to remember the technical doldrums of the 1950s and 1960s, when grainy, scratchy black and white 16mm film stories seemed a permanent way of life. That era slips further and further into the murky past, especially now with the industry preparing for still another technological revolution. This one is called **HDTV, high-definition television.** It provides the clearest, largest, most beautiful TV pictures ever made, and by the year 2000 it should be commonplace in the 90 million or so American homes that have television.

Japan has a clean beat on developing the process. Scientists there started working on HDTV 20 years ago. In the United States, the Defense Department, which needs HDTV for certain weapons systems, has been coaxing tardy American companies into the market with financial incentives. However, Japan has already won the race, resulting in a lot of red faces in U.S. research and development circles, particularly since HDTV will be a $250 billion business within the decade.

TV pictures improve with an increase in the number of lines that make up the video picture, just as increasing the dots in a still photograph enhances that image. U.S. TV is based on 525 scanning lines, while European TV is slightly better, with 625. However, Japanese HDTV goes all the way to 1125 lines, creating the sharpest TV picture ever.

But this blessing comes with a steep price tag. First, viewers will have to get rid of their old TV sets and buy new, more expensive ones—in the $4000 range. Of course, as with innovations of the past, the price may eventually become more affordable. Second, TV stations face a major retooling job. The estimated cost for a typical station to become HDTV ready is $37 million.

TECHNOLOGY IN JAPAN

HDTV is still more evidence of how the Japanese have revolutionized television technology. We already use their cameras, videotape, tape decks, editing consoles—virtually every technical aspect of TV production including the television monitors themselves.

Japan then, obviously, has to be the international leader in TV news high tech. Wrong! In some ways, Japan is at the level the United States was in the 1950s. The biggest irony is that Japan, the country that is always in trouble for dumping cheap computers on America, doesn't have computers in its own newsrooms. Walk through the Tokyo studios of NHK and you'll find reporters writing news stories by hand, *without* computers, *without* even typewriters, electric or manual. NHK, the government network, stands for Nippon Hoso Kyokai, which translates to Japan Broadcasting Corporation. It is Japan's undisputed leader in the availablity of news, particularly at the local level. Sadly, the Japanese, who have mastered the electronic chip to control the largest jets and tiniest instruments, have yet to lick their own alphabet, which is composed of thousands of characters.

The Tokyo Broadcasting System, Japan's biggest commercial network, suffers for it but hopes for the best. TBS sees word processors in "the not very distant future."

And back at NHK, expanding news production does not necessarily mean getting more people. NHK's morning newscast was recently expanded by 45 minutes, its evening show by an hour, and its pride and joy, local "News Center 9," by 50 minutes. Asked how many more people will be added, "News Center 9" producer Kenichiro Iwamoto laughed, then said, "They're cutting people."

An NHK 22-year veteran newscaster in Tokyo, Yoshihiko Kubota, admits that Japan lags behind the United States in electronic news gathering, and that this affects the quality of broadcast journalism in the country. He observes that U.S. anchors read the story **lead**, which is the first paragraph and includes the main facts, then switch to a reporter in the field. At NHK, Kubota, in most instances, reads the complete story himself in the studio because of this lack in ENG.

TUF-TV is a young station with a young newsroom. The TBS affiliate, in the Fukishima prefecture 150 miles north of Tokyo, is only four years old, the people who work in the news operation have an average age of 27, and most of the equipment is brand new. Yet producers use a visual device dating back to U.S. broadcast news in the 1950s. TUF, which serves a population of 150,000 people, uses the ancient **Telop**, a nonelectronic card-type technique for projecting still pictures.

Another small operation, NHK's Shizuoka operation 75 miles southwest of Tokyo, has a staff of only 20 to put together a 20-minute newscast at 7:30 A.M.; a 5-minute show at 12:15 P.M.; then their main cast, a half hour at 6:30 P.M.; and another 15 minutes at 8:45 P.M. As a result, Kazuo Hori, in addition to his vice presidential duties, also writes, rewrites, outlines, and even helps with research.

EUROPE BY SATELLITE

In Europe, there is a satellite revolution, but so far the emphasis is on providing commercial programming rather than SNG. Within five years, viewers will have available more than a hundred satellite and cable channels. Satellite dishes are becoming smaller and cheaper overseas, and expansion is the name of the game. While the heavy emphasis is on entertainment, there are channels specializing in news and sports. Publisher Rupert Murdoch, for example, has launched Sky Television, a nine-channel service costing in the hundreds of millions of dollars.

News operations throughout the continent would like to see some of that money and technology. Italy, for example, also has hundreds of local channels, but news budgets are skimpy. The government's main network, Radio Audizione Italia, or RAI, has three channels and each is controlled by different political parties. TG-1 is influenced by the Christian Democrats, TG-2 by the Socialists, and TG-3 by the Communists. To really complicate matters at RAI, there are no producers, writers, or unit managers. That means the reporter has to do it all, including the budgeting aspect of the story.

So who's in charge? According to Vittorio Panchetti, news chief for RAI's Channel 2, "Decisions are made by committee and many, many meetings every day."

Swedish TV is substantially behind America in live capability. Ivan Nohren, head of Technical Operations for Swedish TV, says it takes at least an hour to get live coverage because of microwave and satellite shortcomings. As a result, 95 percent of Sweden's news coverage is taped and biked back to the studio. The problem really hit home when the country's biggest story in decades hit on February 28, 1986, the night Prime Minister Olaf Palme was assassinated. Nohren says it took at least four hours to get a live picture from the nearby murder scene in Stockholm. So far Sweden has no satellite of its own, but it shares the Nordic Television Satellite with Norway and Finland.

French broadcast journalists do not have full use of a satellite either, although they have several channels that provide news. French viewers choose among Channel 1, which used to be state-owned but has been bought by private interests; Channel 2, the major state-owned outlet; Channel 3, the state-owned source of regional coverage; and Channel 4, which is basically entertainment pay TV with only a handful of three-minute newscasts. Channel 2, or Antenne 2 as it is known in France, has a staff of 440 and a total of 48 electronic news-gathering units, but no satellite capability of its own.

The two giants in British broadcast journalism are the British Broadcasting Corporation, whose programming is funded by license fees paid by TV set owners, and Independent Television. ITV's news arm is Independent Television News. The four basic channels in Great Britain are: BBC Channel 1, a general, popular network; BBC Channel 2, which offers a different, slightly more intellectual product; ITV's Channel 3, the network fed by Independent Television News; and Channel 4, a cultural channel owned by a compendium of independent companies and also serviced by ITN.

Only the BBC and ITN network operations at their London headquarters have impressive staffing and high tech. As one ventures into the field, the staffs and the state of the art tools shrink. ITV's Thames Television in London covers greater London on programs starting at 9:25 A.M. and finishing at 10:30 P.M. Yet Barrie Sales, director of news and current affairs, has a staff of only 40 people, including 12 reporters.

Central Television News in Birmingham is one of five designated network companies producing the bulk of the programming for independent Channel 3. Editorial manager Laurence Upshon covers a quarter of England, a population of 9 million, with a staff of 160.

Rick Thompson has additional problems in the BBC's Midland region. As head of news and current affairs, he has to worry about radio as well as TV. Since the BBC wants to offer a total integrated service, Thompson's 10 newsrooms include eight local radio station newsrooms and two TV newsrooms. The various newsrooms share information through a computerized Telex system. In the Midland News Exchange, stories originating in any of the 10 bureaus are distributed by computer to the 10 newsrooms and London. The result, hard work, but in addition, exciting, complete coverage for an area of 8.5 million people stretching across England from coast to coast.

Glasgow's David Scott has 60 staffers to cover a market of 3.5 million people across Scotland's center. Scottish TV doesn't have a satellite news-gathering system either, but news controller Scott anticipates use of a co-op bird in the next few years.

PROJECTION

For television journalists in other countries, it's not exactly a small world. Many of their coverage areas are considerably larger than those of their counterparts in America. But they accomplish near miracles with innovation, enthusiasm, dedication, quality, integrity, and hope for some of the toys Santa has already dropped on the United States. With satellite programming exploding overseas, however, broadcast journalism in foreign countries is likely to catch up and possibly even pass us in the next decade.

In the meantime, the American news producers will probably mature in their own use of electronics. It took time for management to grow past the novelty of color film. It will also take time for some broadcasters to develop a more logical, more realistic, more journalistic *modus operandi* regarding the blessing of being able to go live from anywhere at any time by microwave and satellite. As it has in the past, at some point journalistic instinct will preempt upper management's tendency to go **live** and around the world merely to amortize the cost of the equipment and to help ratings.

It is also likely that television will suffer through and grow out of the same adolescent cycle with any future technology, such as high-definition television. Stations, for example, are already worried about the conversion costs of this technology, and news budgets, as well as taste, could suffer.

And, of course, there are still a lot of bells and whistles in front of us. It is conceivable, in the next century, that broadcast journalists will dust off **Teletext** and **Videotex**, processes that provide direct two-way communication between viewer and station. Visionaries like KSL's Kinard hope to see news constantly available, cafeteria style. When viewers wants certain stories, they will access them directly into their homes through control boxes.

Newsroom computers in the very near future will also have the capability of projecting video right at the workstation within the newsroom or at a reporter's location in the field. That means that all the key points—studio, newsroom and scene of the story—will be able to send and receive visual materials. All this *and* contented cost cutters! Because of the audio-video-print communication ability of computers, news departments will be able to send fewer staffers to the point of origination, no matter where that may be.

Despite competition from fiber optics, nobody will get away with shooting down the satellites. If anything, the United States will go in the same direction as Europe and end up with a multitude of channels. For American viewers, that will mean a feast of entertainment programming as well as news. Finally, we will be able to tune into newscasts from New York as well as from Duluth, from Moscow as well as from Sacramento.

Again, the industry and its audience will go through the gee-whiz, trendy, let's-play-it-to-death syndrome before settling down. But it will settle down, because solid journalism isn't that fragile. If the message, and its messengers, could survive the spoken word, the written word, radio, black and white film, color film, and videotape, they can certainly handle, and continue growing, "live and direct."

2

---◼---

LOCAL-LOCAL, LOCAL, AND BEYOND

The Ever-Widening Parameters of TV News Coverage

Local TV—armed with the high-tech equipment, a profit motive, and severe network slippage—is redefining its turf. Anybody goes anywhere to cover anything, but who's minding City Hall?

Little did the Russians know back in October 1957, when they launched **Sputnik**, the world's first man-made satellite, that it would be the local news directors of the United States, not the Bolsheviks, who would take the ensuing space technology and conquer the world.

Yes, tomorrow is here and the world belongs to the news directors. From coast to coast and from the Canadian to the Mexican borders, TV journalists are redefining local news. And that new definition contains no geographical boundaries or subject restrictions. International summit meetings anywhere, earthquakes in Mexico, any war from Nicaragua to Vietnam, papal tours, presidential trips to foreign countries, national political conventions, the Superbowl, and angles to local stories on any continent now grace the assignment board. The obvious concern is how many other stories, from City Hall around the corner to that dangerous intersection up the street, are being wiped off the board as a result.

The presatellite definition of local news entailed covering the station's immediate market area. For smaller stations, it meant staying within their towns. Larger news operations, and the more conscientious ones at that, might include the state capital in their beats. But going outside the state to cover a story was rare, as was going overseas. Covering news of national and international importance was clearly the network's domain.

The big breaking stories of the 1960s—America's civil rights revolution, the Vietnam War, and the assassinations—created a generation of **news junkies**. In the 1970s, when stations realized there was money to be made in local news, the field exploded into a cacophony of Action News and Eyewitness News formats. Local newscasts got longer and went on earlier, well before the network news, so the natural step was to add videotaped packages of national and international news to the casts. With the ENG and SNG technology of the 1980s it became economically and technically feasible for the stations to jump the last hurdles, time and distance. They could now personalize national and international stories with their own reporters, anchors, and call letters.

On the surface this trend appears to be negative and self-serving and at the expense of local coverage. However, broadcast journalism is hardly a profession of absolutes. Spectacularly professional, exciting, and meaningful journalism is still taking place out there, thanks to high tech. And a lot of people in the business today have the same standards the industry did 10, 20, 30 years ago. The same type of person who could produce informative *and* entertaining pieces with the Stone Age tools of yesterday knows how to do it equally well with son of Sputnik today. Of course, for some news directors today technologies are little more than toys. But there's an unfortunate precedent for that. Generations ago similar news executives also played at broadcast journalism with the new toys and tools of their particular era.

THE RATINGS SYNDROME

A major difference between the past and today is the contemporary TV journalists' addiction to the numbers game. Whenever the subject of news coverage and program content comes up, that of sweeps and ratings isn't far behind. The **sweep** is a month-long, four-times-a-year ritual during which the two ratings services, **A.C. Nielsen** and **Arbitron**, survey viewers to determine which programs are the most popular. The cost of the commercial minute sold on the news is determined by the results of the sweeps. A difference in rankings and rating points means a difference of millions of dollars in income.

Tell today's news directors about the good old days of the 1950s and 1960s, when producers were told simply to do the best possible job and forget about ratings, and watch their eyes glaze over. They become somewhat like John Steinbeck's Lennie in *Of Mice and Men*, begging George to describe that farm they'll never get. At NBC News back in the 1960s, prime time entertainment programs were **preempted** constantly for instant news specials or special reports on major stories. Audience size did not become a criterion until the decade neared its end.

But the hard cold reality of today is that station owners demand ratings, which they cash in for money. Many stations post the overnight ratings comparisons in full view of the entire news department. That competition is one reason for the emphasis on razzle-dazzle, for news crews from Hometown, U.S.A., traipsing worldwide to show the call letters.

TV NEWS GLOBALISM

The genuine pursuit of a local angle to a national or international story makes journalistic sense. It is also logical to dispatch a team if the network can't or won't. John Spain, news director of WBRZ in Baton Rouge, Louisiana, is all in favor of going overseas if the reporter can show an impact on the community. Spain, like many news leaders, says the stations should have been doing that all along. However, the ABC affiliate news director is vehemently against going for the sake of going.

Phil Balboni, news vice president at WCVB-TV, Boston, thinks it's a fine idea to hit the road if the stations supplement, not imitate or duplicate, network coverage.

However, there are some dangers to covering the news worldwide. For example, influential forces, such as consultants, often push stations into spreading their wings—but without good reason. Advice like this may be questionable, particularly when pushing the stations into globalism creates homogenization of newscasts.

The most impressive evidence of globalism is in *Broadcasting*'s annual roundup of local accomplishments, which are submitted by the news departments themselves. This magazine, in its 1988 survey, heralds the news with the words, "Local journalism is fast becoming a misnomer. Local journalism is regional journalism is national journalism, with satellite technology and networking enlarging the reach of local stations by minimizing the constraints of time and distance in the pursuit of a story." The magazine, for 29 pages, chronicles the accomplishments of more than a hundred news operations. Some examples:

- Eau Claire, Wisconsin, does a report on supercomputers from California.
- Washington, D.C., goes to Pennsylvania for a murder story.
- Scranton, Pennsylvania, reports from Wall Street.
- Wichita, Kansas, goes around the world on the tail of a Boeing 747, setting a speed record.
- Anniston, Alabama, accompanies a local cerebral palsy group to Orlando, Florida.
- San Francisco tours the 38th parallel in Korea.
- Miami chases a hurricane to Jamaica, the Cayman Islands, Mexico, and three cities in Texas.
- Jefferson City, Missouri, originates its entire newscast each night during three weeks on the road from different locations, including a barge, a construction site, a hospital, a truck stop, and a fire department.
- Tampa–St. Petersburg, Florida, follows the National Guard to Nicaragua and Panama.
- Davenport, Iowa, does its newscast live from a riverboat on the Mississippi.
- Buffalo, New York, broadcasts a newscast live from a new baseball stadium.

A CASUALTY: STATE COVERAGE

In California, local news directors often dispatch reporters to Asia, Europe, Africa, and throughout the Americas. Yet the NBC affiliate in

San Francisco closed its office in Sacramento, leaving the state capital with no bureaus from out-of-town TV news stations. Just 20 years ago, the major TV stations in Los Angeles, San Francisco, and San Diego had full-time news bureaus in the capital of a state large enough to rank as a nation. The *Los Angeles Times* felt strongly enough about the desertion to run an editorial pointing out that California's 28 million residents rely on television as their primary news source. Acknowledging cost as the reason for the stations' exodus, the *Times* chided the networks because they still manage "to maintain bureaus of more than 100 people each in Washington, but their local offspring decline to spend enough to keep a single correspondent in Sacramento."

Other states aren't doing much better. There is only one full-time TV bureau covering state legislatures in New York and Massachusetts.

Ironically, the new technology has converted Washington, D.C., not the state capitals, into a local **beat**, or coverage area, for many stations throughout the country. The networks, satellite services, and regional groups provide customized reports for local newscasts from the nation's capital. If the story is big enough, news directors will send their own teams.

Jim Snyder, vice president of news for the Post-Newsweek stations since 1969, has no problem with globalism as long as it is logical and journalistic. His news authority covers NBC affiliate WDIV in Detroit; CBS affiliates WFSB in Hartford, Connecticut, and WJXT in Jacksonville, Florida; and ABC affiliate WPLG in Miami. Any one of those news departments will leave town for the right reasons. Snyder thinks in global terms for his local stations despite being an old-timer who started as a writer at the nation's first commercial radio station, KDKA, in Pittsburgh. His news people originate from national political conventions "in order for Post-Newsweek stations to talk to their anchors there and talk live with members of their delegations." Cover the pope in America? "You bet, particularly in view of the two million or so Catholics in the Detroit area."

THE LOCAL-LOCAL CONCEPT

Ernie Schultz of the Radio-Television News Director's Association is a believer in local news, but he doesn't mind stretching the definition. Could it stretch clear to the revolution in the Philippines? "Absolutely, once it all shakes down. If there's a local angle they (the news directors) will find it."

Yet the news veteran sees a trend to even more localized news, a concept appropriately labeled **local-local**. Market research has shown that people want local-local, says Schultz. "I mean it's not just enough to say that you're covering Sacramento. You cover what happened at the First Baptist Church in Sacramento. Local-local."

One reason WCCO-TV's number one–rated news operation in Minneapolis has a consultant, Audience Research and Development of Dallas, is because it specializes in local-local. Outside consultants, which will be discussed more thoroughly in Chapter 4, are hired by station management to provide critiques and guidance for newsroom operations.

The CBS affiliate's half-hour program at 4:30 P.M., "Newsday," is evidence of the approach. Emphasis is completely local and often is produced from a remote location. "Newsday" also has an audience call-in capability. The station's 5 P.M. half-hour lead-in to the network news is 80 percent local. The half-hour following the network news is 85 percent local and 15 percent regional. The 10 P.M. show is at least 60 percent local-regional and the rest is national-international. The noon news, however, is heavily non-local, with a good deal of material coming from CNN.

Video Nomadism

While Reid Johnson is dedicated to local-local, he's also into video nomadism in a big way. WCCO cameras have panned such far away locations as India, Vietnam, Cambodia, Thailand, Ethiopia, and Central America. Johnson will commit outside Minnesota or the United States if the story has a Minnesota tie or a different perspective. In Vietnam, the Minneapolis station tracked Minnesota military personnel. In India, it followed a businessman who took the money he made in Minnesota back to his homeland to build a hospital.

The policy of this Twin Cities station is an example of how easy it is to cross that thin line from legitimate to questionable coverage. WCCO had no local motive for covering Ethiopia and Central America. In Ethiopia, the station went after "a different perspective," finding "real people and faces behind the cold statistics of the starving." The news director justifies covering Central America by maintaining, "We only see it in one-, two-, three-minute form. We really want to take a much longer look at it and try to tell the story as completely as we can."

WCCO competitor KSTP-TV also claims to be on top of City Hall, but it too spends significant time outside the Twin Cities. To Larry Price, the Ku-band truck means an extended coverage area and the opportunity for electronic show and tell. This ABC affiliate, like the many stations in the *Broadcasting* survey, likes to take its newscast out of its Minneapolis studios and originate on the road. The station calls itself "Minnesota's news champion," its coverage stretching well beyond Minneapolis-St. Paul and the suburbs to all of Minnesota and parts of Wisconsin, South Dakota, and Iowa.

Does the emphasis on satellite transmission black out KSTP's or WCCO's backyard? Price at KSTP insists his reporters go to City Hall regularly and stay in touch with the police, in addition to other things they cover. WCCO claims the only difference a satellite makes is that they now tell those local stories better.

KGO-TV's Harry Fuller makes no bones about it. As far as his San Francisco operation is concerned, "anything that happens in *northern California* is local-local." According to the news director, 50 percent of his 5 P.M. newscast is local-local, as is two-thirds of the six o'clock news. The 11 P.M. varies from a third to two-thirds local-local.

Actually, for KGO, anything happening anywhere is fair game. Fuller's criteria are simple. Pick topics and issues that are very relevant to the market and don't do the same thing everyone else is doing. His third, possibly most meaningful proviso, is to move on a story "if it is so important or such a compelling story to the local market that I can't afford not to be there competitively."

This approach opens the door to still another form of **rat pack journalism**, that tendency for everybody to cover the same story the same way. The practice of having huge numbers of local TV staffs cover national events, such as inaugurations, and international events, such as summit meetings or presidential treks overseas, is questionable. The likelihood is slim that reporters who lack the expertise, contacts, and know-how in this arena will come up with a **beat** (trade talk for an exclusive story) or even a meaningful report. The money spent here would better go for a variety of other needs, from local investigative reporting to opening up new coverage areas.

Yet, it makes good journalistic sense to dispatch crews over great distances to cover stories with a direct impact on the community. This is particularly valid for cities heavily populated by people from the area affected. For example, KGO covers earthquakes in Mexico because the Bay area has a large Mexican-American population, residents vacation in Mexico, and San Francisco is quake conscious.

THE PROMOTION FACTOR

But why send crews all the way to Miami to cover the arrival of the pope? KGO's Harry Fuller gives an honest, to-the-point response. The

Catholic leader was en route to the first papal visit to the Bay area, and KGO saw a way of promoting its own saturation coverage.

This is the own-the-story syndrome, a delightful situation in which the audience watches you from the very beginning, when the story first develops, and stays with you as it unfolds. Competitively, it can mean winning or losing at sweeps time.

Would Fuller have sent news crews to Hong Kong to blatantly hitchhike off a network entertainment series? That's what Los Angeles's KNBC did with "Noble House." The station sent a reporter and crew to do a string of stories from Hong Kong to air locally and jibe with the network miniseries. There was no local connection or justification, other than mutual promotion. Fuller admits he would have been tempted. "The whole cross-promotion business in television is what's happening. If you have shows in your prime time during ratings periods that deliver an audience, do whatever you can to get a portion of those people to stay up to at least 11:15 (P.M.) so you can get them for the first quarter hour of the newscast. So people will do damned near anything."

He also verbalizes what many news directors have caught in their craws. Fuller's going to do what he has to "because if I won't do it, they'll fire me and find a news director who will." Unfortunately, Fuller is right. Journalistic autonomy in the local station, or even in the network, has become rare since news has become a profit center. The produce-profits-or-get-out attitude is prevalent in modern society, whether we're talking widgets or news.

Herb Dudnick's KRON crews don't wander as much overseas, but his reason is not necessarily altruistic. The NBC affiliate's problem is that it is number three in the ratings, and that's the basic reason, according to Dudnick, for sticking to the San Francisco Bay area. If KRON-TV were number one and established in the local market the philosophy would be different. So, for Dudnick, the priority is to cover the local area first and start moving out later.

When he does leave town, the news director bases decisions on the importance of the story to the local area and the cost and impact on other local coverage. Soldiers sent from California's Fort Ord, which is some two hours drive from San Francisco, to Central America became a local story for KRON. But the station used a less convincing rationale for covering a U.S.–USSR summit meeting in Washington, D.C. Dudnick calls that coverage "an unusual occurrence"; he says he did it because he wanted to involve two of his anchors who have "network, international experience" and "because I had a bureau there."

The news executive admits covering stories away from home takes away from local coverage. "You just weigh it. It's no different than everything else every day in terms of producing a story." When KRON does move out, it will do it in a big way, establishing bureaus in the Pacific Rim and Central America.

Dudnick considers the 5 o'clock newscast far enough away from the network show to warrant a non-local identity. He describes the cast as an opening tease to the two-hour block that includes the network's "Nightly News." His 5 P.M. show is designed to give a broad idea of what's been going on nationally, internationally, and locally. The 6 o'clock news is "about 90 percent local, but remember we consider Honduras to be local and we consider Panama to be local. We consider the Philippines to be local." San Francisco's demographic profile includes substantial populations from those countries.

For a non-local story to make KING-TV's 5 P.M. newscast, which hits Seattle air an hour before the net's "Nightly News," it has to be "compelling," according to Donald Varyu. He says that 80 percent of the NBC affiliate's 5 o'clock is local-local, 5 to 10 percent is regional, and less than 10 percent is national and international.

Stories that drive KING crews outside the market area are "basically issue-oriented." However, KING, like many of the stations visited, will go out of state to cover stories it feels will not receive coverage on the network.

The station often does exemplary journalism. When the home-based Boeing Company–built Aloha Airlines jet crash-landed in Hawaii, KING crews fanned out to Hawaii, Washington, D.C., and New York. It's the kind of story over which Varyu will break all sorts of rules, including length. The sensational Aloha Airlines story, featuring the jet's dramatically shorn-off top, ran as long as seven minutes on KING air.

The Pacific Rim is also important to the King Corporation's flag-ship. The station has done series on the economic impact of Korea, the Olympics, and local involvement in Nicaragua. The most frequently visited overseas location for KING-TV crews is the Soviet Union. Motivated by a "peace movement, which exists in the power structure of the city," KING initiated a series of live televised programs directly linking the United States and Russia.

As far as KOMO's James L. Boyer is concerned, Seattle is only part of the coverage area. The news director considers his news coverage area to extend from the Canadian border down almost to Portland and from the Pacific Ocean to the Cascades. He sees his concern for such nearby cities as Bellevue, Everett, and Tacoma as equal to his concern for Seattle, his home base. Seattle city hall? That's not a beat for KOMO reporters. "The city of Seattle is just one more place and I don't think any of our cities is more important than anybody else's." According to Boyer, the emphasis is on covering issues and stories, as opposed to geographical locations.

To Boyer, a San Diego, California, angle on the Green River murders, a sensational ongoing serial killing case in the Northwest, warrants leaving town. So do the national political conventions.

KOMO also sent one of its anchors, a newcomer with Washington, D.C., experience, to the Reagan–Gorbachev summit in the U.S. capital. The rationale was weak and there was no local angle, but that's not unusual for many stations. Boyer, who was banking on the anchor's experience as a reporter, admits, "It was also highly visible, a big ticket story that allowed us to showcase him."

KOMO cameras have also been to Russia, but not as often as KING's, and to California for earthquake coverage. Boyer's description of KOMO's 5 o'clock cast doesn't differ substantially from KING's in philosophy. He describes it as a local newscast, "but we will cover state and national news if it is extraordinarily big and can't wait until Peter Jennings at 6 P.M."

A typical July 1988 cast started with KOMO's reporter live in San Diego with a tie-in to the regional Green River serial killings, followed by stories on:

- a local prison escapee
- a local crane accident
- a local fire
- a ship, loaded with dope, seized in the local port
- local seniors banded together to fight crime
- two consumer subjects

About halfway through the newscast it turned to national news for about ten minutes and then presented the usual sports and weather.

KOMO's newscast following the network news at 6:30 is all local. It includes a truncated version of local news of the day and a daily "Money Watch" segment featuring area businesses and economy.

WNEV-TV in Boston claims a true emphasis on local and regional news, but staffers will travel anywhere in the world to cover a story with a related angle. The CBS affiliate's crews saddle up "when it is logical." For example, if Massachusetts governor Michael Dukakis is doing something exciting in another state, WNEV will figure out a way to deliver it live in Boston.

WNEV has gone to Mexico, tracking a local Catholic church official who was involved in earthquake rescue operations. However, the
reason for sending an anchor to the Iran–Iraq war front was strictly
exploitative. The broadcaster had Canadian citizenship and Jeff Rosser saw it as an opportunity to cover the war with a local face.
Newspaper critics accused the Boston station of showing off the anchor doing something that should have been reserved for a network.
But the viewers had a very positive reaction, insists Rosser, who
maintains they appreciated the commitment and the personal approach. The Boston station wouldn't go to the Philippines for the
revolution, but half the stories the satellite truck rolls on are, nevertheless, out of state.

While WNEV's one-hour newscast at 5 P.M. averages 95 percent
local, that percentage can drop by at least 15 points if something major
happens overseas, with the remainder of the newscast going to the
breaking story. The station fills out the story with interviews with
New England congressional representatives or senators in Washington through the Potomac News Service, local and regional reactions,
and footage from the network syndication service.

ONE-STOP NEWS SHOPPING

One-stop news shopping is an approach that fills a vacuum left by
networks no longer in the instant news specials business. There was
a time when the nets would quickly move in on a major story, cancel
lucrative entertainment programming, and provide thorough coverage. But that is not the case in today's budget conscious environment.
And, although CNN does provide outstanding news programming on
breaking stories throughout the world, its programming doesn't include angles related to the viewer's hometown.

Because of their networks' abdication in a vital coverage area,
affiliates are turning back the clock and reviving instant news specials.
The result is like one-stop supermarket shopping. If viewers can get
everything they need—local as well as national and international news—
they'll continue deserting the network for the coverage and the faces
of their favorite, familiar anchors and reporters.

The reality, according to Jerome Nachman, is that "people won't
wait until 7 o'clock or 6:30 to learn something they can learn at 4 or 5
o'clock." Nachman is the former vice president and general manager
of the NBC owned and operated WRC-TV in Washington, D.C.

Nachman sees nothing wrong with sending local reporters to

cover earthquakes in Mexico. The idea, he says, is to find a local angle and get someone there. With this formula, any dateline can be a local story. "You know, in New York, when I'm serving an ethnic community that large, going to Israel is a local story." The outspoken proponent of strong affiliate control was news director of WNBC-TV in New York prior to his stint in Washington.

Some local news executives tend to bristle when asked whether local news is suffering from the new expanded look in coverage. One reason for the testiness is the barrage of criticism by former network executives and the print media. With their newsrooms expanding, budgets ballooning, and profits growing, news directors tend to be a bit arrogant and defensive on the subject. Nachman, for example, was asked about the impact of expanded national and international news content on local in-depth coverage. His reply: "Why are you giving me a death . . . scenario? News has always been news and always will be. This is what you've got. This is what happened. This is where you go. This is what you miss. Just like the local bureaus in New Jersey getting short shrift when the Hindenburg blew up."

Gary Mehalik insists that mobility enhances local coverage for WTVJ in Miami. He looks forward to the upcoming linkup between Conus Communications, the satellite-fed mini-network, and Japan. When the linkup is made, Mehalik will dispatch his crews to originate stories in Japan and send them back via the Conus operation. He also wants similar linkups with Europe and South America. Even without the satellite service the Miami station has sent reporters to cover the Philippines, the Vatican, El Salvador, and Cuba.

Jim West of Tampa station WTVT is comfortable with his coverage outside the city. His is among the stations for which Washington, D.C., is a local beat. WTVT has a Washington-based freelance reporter who provides personalized angles for Tampa viewers. "If a local mayor is making a pitch for federal revenue-sharing funds in front of a House or Senate committee, that's not national, obviously, but if we want it, we cover it."

Although West's staffers will cover earthquakes in Mexico if they can find a local angle, they hesitate sending crews further afield, such as to Rome for preliminary coverage of a papal visit or to Korea for a pre-Olympics look. However, when Floridians, concerned about terrorism, were afraid to take European vacations, WTVT did reports from London.

"CBS Evening News" hits the air at 7 P.M. in Tampa, which means WTVT producers lean toward using national or international stories at the top of its 6 o'clock newscast.

Despite the boom, TV news staffs are still small compared to those of newspapers. As a result, most stations can't afford to assign reporters to specific permanent beats, such as fire, police, education, or city hall. Because money is limited, news directors have to decide story by story how to expend it. This places West among the many broadcast journalists who won't cover a meeting at city hall just for the sake of it. WTVT decides on the basis of how many people are affected.

Despite this policy, the community is vital to the CBS affiliate, and the station recognizes this. WTVT's news executives meet regularly with community leaders. The importance of the community is being realized by more and more stations nationwide.

West's competitor, Ken Middleton of ABC's WTSP-TV in St. Petersburg, has a more conservative coverage policy. Middleton dispatches crews outside Tampa Bay only when there is a local tie-in. Sending a crew to the Philippines would be out of the question. However, sending a crew to Central America would be acceptable if National Guardsmen from Tampa Bay were involved or citrus farmers from Florida were beginning to grow oranges there.

NETWORK ATTITUDES

Presumably, the network syndication service would agree with WTSP's Ken Middleton and discourage country hopping. But to NBC's Jo Moring, whose job it is to keep the affiliates happy, this trend of local stations adopting a global news budget is a positive one. The syndication executive says that that means affiliates are going with the most important story, regardless of origin.

Moring's feelings are indicative of the network's evolving attitude of: If you can't beat them, join them. Sanctioning the local stations' broader coverage base opens the door to a brave new relationship between the network and the affiliates.

Paul Dolan, on the other hand, who is an ABC syndication executive, sees danger signs. It's possible, he warns, for the Buck Rogers hardware to have a negative impact on the local station's primary mission of covering local news. Nevertheless, he recognizes that there are checks and balances: "There's enough competition that, if a station is losing touch with its community, other stations will use that to show the weakness of the station."

TRAVELING WITH DISCRETION

KGW-TV of Portland, Oregon, is among those stations that need a valid story hook before hitting the road. The NBC affiliate prefers a direct local angle, but if it's an issue of general importance for the Portland area, KGW will pack up and go. For example, news director Reagan Ramsey sent his people to Vietnam on the 20th anniversary of the Tet offensive. A large number of Vietnam veterans live in Portland, and the war had a major impact on the community.

Because he moved to news from the production area, Ramsey learned journalism on the job. Yet, he sounds like a broadcast journalism traditionalist of the old school. His crews will not go to Nicaragua, the Middle East, or to summit meetings to do instant analysis "because we don't have the expertise." The news director would show his flag in international trouble spots such as Israeli occupied territories only if, for example, doctors from the Northwest were going there to treat refugees. As for going to Central America to cover a revolution, even "network crews can't tell you what's going on in Nicaragua and they have crews that are supposedly down there on a daily basis."

According to Ramsey, KGW's commitment to local and regional coverage is greater than it's ever been.

Salt Lake City's KSL-TV and Helena, Montana's KTVH differ in size (KSL's market is 40th; KTVH's is 209th) but share local news philosophy. KSL's J. Spencer Kinard and KTVH's Ian Marquand are both seriously committed to local news. Both see dangers in the national trend toward Jo Moring's global news budget.

Kinard's producers don't even think about cannibalizing the network's evening show in their exclusively local 6 P.M. newscast. However, KSL does have a full-time Washington bureau that extensively covers Utah lawmakers in the nation's capital. In addition, the station will originate newscasts from national political conventions, concentrating there on the Utah delegation. KSL will send teams anywhere to follow up a Utah angle. That could include trips to Australia and New Zealand for a tour by the Mormon Tabernacle Choir or any war involving local soldiers. What Kinard won't do is cover the collapse of Saigon, summit meetings in Moscow, or Nicaragua.

KSL's guidelines for leaving the state are simple: "If we have to do it, we do it." Assignment editors are trained to know that a murder involving a polygamist clan in Houston, Texas, is worth moving on because of local interest in polygamy. So is the pope in Phoenix, Los Angeles, San Francisco, and Fresno. Kinard notes, "We're owned by the Mormon Church and some people think that's all we care about. I thought it was important to show them we cover the news in . . . religion for all people."

Yet Kinard is one of the few news directors to recognize as well the many wrongs of local TV news. "The things I see are too much shallowness, too much going to the summit in Moscow or the summit in Washington, even to the convention in Atlanta, just to be there. And to spend all the time and money, and it's just showing the flag."

What's wrong, Kinard continues, is "too many times of going live just because we can go live and we think that's going to impress the public. Too many times of standing in front of an empty room while an hour ago there was a big crowd here and they're all gone now and I'm here."

Kinard doesn't let his own station off the hook. "We do it, the others do it, and I think it's done too much. We've let the technology, this wonderful technology, dominate too many of the things we do for the wrong reasons."

If Ian Marquand, a strongly committed local news broadcast journalist, had Kinard's budget, would his philosophy about covering news be different? He tries to limit out-of-city trips to a 50-mile radius, but cost is not his only reason. KTVH's criterion is, simply, to cover whatever is of interest to the local people. For example, any story involving economic growth is important to Helena residents, who live in a state that is economically depressed. That means any story about the revitalization of the mining industry gets the gears working.

Of course, Marquand's coverage has to be affordable, but above all, he won't overlook what's happening in Helena to go out of town. "It has to relate to the general viewing area and it has to fit into our ability to cover it and still get a newscast on that night." In Helena, that translates into good, solid news judgment, just partially fashioned by the availability of money.

LOCAL-LOCAL OVERSEAS

With specific channels dedicated to local news, complete, frequent, and meaningful local news coverage is virtually guaranteed in many countries abroad. And at least so far, there is little evidence of the characteristically American fuzziness and bickering in the relationship between network and affiliate.

However, in some countries, this situation may not last for long. The Japanese, for example, look wistfully at America's commercial success and could be tempted to put the rating point before journalistic integrity.

In Great Britain, on the other hand, there will always be a BBC, or the **Beeb** as it's affectionately called, and ITV. This means that from the tip of Wales to the Scottish Highlands and over to Ireland, audiences will continue to see the finest in national *and* local TV news. This is despite the fact that the 15 regions the BBC territory is divided into have a primary responsibility of servicing the network. These 15 stations are, in a sense, wholly owned affiliates, and if the home office says to drop everything and cover that political debate up the road, they do it. The line between network and station is clearly drawn in Britain, unlike the tug of war in the United States. Yet, the network's dominance has little if any negative impact on the regions' ability to cover their own local news.

ITV is more liberal—its regions are not required to provide coverage for the parent company. Instead, they are permitted to sell it to the news division, ITN. The attitude at Thames Television, the ITV operation in London, is typical of both government and independent operations. Thames leaves all national and international news to ITN, which means it does not cover any non-local story for its newscasts, even if the story breaks in its own backyard of London. That would be comparable to a local news operation in Washington, D.C., ignoring a major story at the White House. Thames news executive Bob Kirk finds the U.S. trend bizarre. He says the coverage of international news is "probably the biggest difference between the two countries. If I were in charge of NBC I'd be very, very worried."

In Sweden the situation is unique. Its localized news service, TV 2, is available in eight different regions and each has a different theme. In Örebro, for example, which is less than 150 miles west of Stockholm, the operation emphasizes culture, which means the head of that region, Jan Hermansson, dedicates most of his programming to subjects related to the arts. Evidence of his specialized theme is seen mostly in non-news programming and in the expanded half-hour newscast on Friday nights, when producers are not devoting that 30 minutes to an investigative report. Hermansson's other newscasts, the nightly 15-minute programs, are solidly local.

Örebro's coverage during a typical September was quite different from that of an American TV station of comparable size. Örebro TV news covered 39 crime stories and *almost the same number of economic stories, 38.* There were 25 environment and nature items. In still another departure from the U.S. emphasis, Örebro had the same number of sports and political stories, 14 each. There were also 11 medical stories, seven stories on education and culture, and three stories each on agriculture and tourism.

The trend toward U.S. TV news customs is strong in Japan. NHK even signed up with the American company Conus for satellite-fed international coverage. Kenichiro Iwamoto considers his local news operation's Conus feed a godsend. When major stories break, such as a plane crash in the United States, the coverage is remarkably similar to American coverage. Conus provides the news producer with the visuals and, often, a reporter from a local TV station for interviews with the NHK anchor.

Actually, Iwamoto claims Japan is outdoing America in its emphasis on foreign news. "The major differences between U.S. and Japanese broadcast journalism are that we cover more foreign news, run longer stories, and have fewer conflicts between the network and local operations."

The Japanese approach to local reporting is far more personalized than even the prototype of reporter involvement, the Eyewitness News format in the United States. Noriko Tanaka's day, for example, begins with the Tokyo reporter finding her own stories without the help of an assignment editor. She also talks to her friends "to find some newsworthy item," she says. After making "a lot of phone calls I meet people, take them out to dinner, and get them to talk about things." Tanaka, who works for TBS's local "Teleport 6" program, then decides which people she wants to interview and what kind of visuals she can get. Finally, and unlike U.S. reporters, Tanaka herself does the physical editing and directs it on the air. Tanaka is a rarity among Japanese broadcast journalists. She is among the very few who studied journalism, a course she took in the United States.

NHK's Masayo Nakajima, a reporter in Shizuoka, goes directly to contacts' homes instead of their offices or the police department. His mission is "to have personal connections with the contacts so they can easily tell me what is going on."

Local newscasts in Japan often concentrate on **soft** rather than **hard news** stories. In most Japanese cities, the man in the family gets home from work late, so the 6 P.M. newscast is blatantly aimed at women. Takeo Takahashi, at TBS's station in Fukishima, acknowledges there is very little in hard news and quite a lot in soft, featury material on his news because of a predominantly female audience. NHK's Kazuo Hori also admits that his 6:30 P.M. newscast is geared that way. "We make the politics very short, like three minutes at the most, because politics is very complicated and not very interesting for the housewives and young people." The opening of a new supermarket, Hori says, is of greater importance for his 6:30 audience in Shizuoka.

Hori also does a lot of local crime news. Despite pressure on NHK and other local Japanese TV news operations to cut down on the coverage of violence, Hori feels crime news is "the most important" for his audience.

PROJECTION

Being considered best in Japan has a lot to do with emulating "Made in the U.S.A." TV news. Now that's a thought that could keep concerned observers of our broadcast journalism awake nights. However, futurists could be resting peacefully. Certain inevitable, and positive, trends seem to be emerging here and overseas. Instead of trying to erase the writing on the wall, critics could be constructively helping to marry the standards and ethics of yesterday's journalism with the pace and breath of today's.

TV news is simply going in the same direction as the daily newspaper, and it's unlikely anyone will stop it. The newspaper provides all the informational goodies—local, state, national, and international news; sports; business; finance; medicine; gardening; human interest; opinion; entertainment; and, even, **muckraking**—in one bundle.

With networks reluctant to spend any significant amounts of money or imagination on fresh news programming, the viewer is faced with huge vacuums in informational needs. Because of time differences, network news arrives in homes stale. Because of budget problems, network news coverage of major breaking stories may not arrive at all. And because of technology, local TV newsrooms are no longer restricted to their market areas.

As a result, none of the news executives interviewed anticipated a retreat from overseas coverage. Most of them expect even more coverage of national and international news on local newscasts in the years ahead. There's no talk about corking up the genie who has given them the technology, bucks, and passports to the world beat.

As local stations fill in news programming holes left by the networks throughout the day, it is inevitable that their newscasts will cover whatever is breaking during these shows. They will be complete newscasts that take full advantage of the increasingly customized network syndication feeds and the other purchasable services such as CNN, Conus, and the Financial News Network.

As for the coverage originated by the stations themselves, the critics and the viewers will see to it that the trend of using a satellite simply because it's there will be a strictly passing one. An impressive barrage of solid journalism is already bouncing off satellites, with stations starting to concentrate seriously on local, meaningful angles overseas.

A vitally important coverage area that will continue shrinking in the immediate future is the one that affects people most directly, their own communities. While local stations will be capable of broadening coverage maturely, the process itself will spread station resources too thin for any dramatic emphasis on strictly local news. Some news departments, such as that of the King Company, are serious in their commitment to safeguard the public's interest at home as well as abroad. But nationally it is an endangered service.

The viewer needs to demand local substance along with the inevitable sports, weather, and traffic. It won't take long for newspaper readers to realize what is missing on local TV news. The newspaper's account of *yesterday's* local news won't satisfy TV viewers' appetites for faster, more complete delivery of events happening in their very backyards.

It is also possible in the years ahead that the networks themselves will help fill the gap by following the European approach and designating certain channels exclusively for local and regional news. The National Broadcasting Company's owned and operated stations are already seriously contemplating a 24-hour local news service on cable channels in the regions. The nation's first service, Bay News Center, is scheduled to begin in San Francisco in spring 1990. The around-the-clock local news programming is a joint venture between NBC's KRON-TV and public TV station KQED. It will be seen on the public broadcaster's co-owned station, KQEC-TV, Channel 32. The potential here is exciting, particularly considering cable's ability for direct interaction with the audience. This interaction could mean three-way access among viewers, journalists, and news sources. A regionalized cable news channel is just one possibility. But it's one that has the ingredients necessary for the continued success of local television news and the actual survival of the parent network news operation: freshness, innovation, excitement, and sound journalism.

3

---□---

IS THE TAIL WAGGING
THE DOG TO DEATH?

Changing Relations Between
Local and Network

The advertising dollar can be stretched just so far. Local news becomes more profitable and independent as network dominance shrinks. Both claim de-tente, but the war may already be over and won.

More than 4000 local television troops crowd into the national political conventions to report them for their home stations. They out-number the network broadcast journalists by about four to one as they cover a story traditionally done by the networks. Similar local news coverage swamps the presidential inauguration in Washington, the Superbowl, plane crashes, and any major breaking story at home or abroad. ABC News airs a one-hour TV special on drug abuse and makes TV news history by tapping 73 affiliates for two-thirds of the show's material. The special, "A Plague Upon the Land," is heralded by the ABC brass as the first in a series of joint projects.

As far as the old-timers, former network television news giants from the top-level executive ranks such as CBS's Fred Friendly and Ernie Leiser, are concerned, the burgeoning affiliates are wagging the dog. And, claim the past movers and shakers, they may also be sounding the death knell for the networks. It's a fear that's echoed as well by not-so-old-timers still in the news business. NBC correspondent Doug Kiker outlined the threat at the November 1988 Radio-Television News Directors Association convention. The network reporter told the news executives, "Local news has become immensely profitable and competitive with the networks in covering major stories." As recently as the 1970s network personnel weren't making concessions like that.

A. C. Nielsen, the survey expert, reports a drop of more than 13 percent in viewership of the three network newscasts in eight years. It is estimated that the networks' overall share of the viewing audience will sink to 60 percent within the decade, a 30 percent loss since 1981. At the same time, the Roper Organization reports more people than ever say their only source of news is TV, which they consider the most credible news source. In other words, more people are watching news, but it's not network news.

The American Broadcasting Company's concern about such findings borders on panic. The network insists that viewers in bars, hotel rooms, second homes, and college campuses be included in the nose count of viewers. ABC officials claim there may be 4 to 10 million uncounted viewers. The Columbia Broadcasting System joined ABC in July 1989 on a joint research project they claim reports close to 6 million college students and working women TV watchers who don't show up in the ratings. According to the networks, 2.6 million women tune in at their workplaces and 3.1 million college students watch away from home.

Viewers mean rating points, which mean more money paid by advertisers per commercial minute, which means profits—and that's what it's all about. The affiliate, like any business, is not going to back off from making money. Nor are the networks, or **webs** as they are referred to in the media journals, going to apply too much pressure. If they do, the stations will simply bolt to another affiliation with a more laissez-faire attitude, thus endangering the very backbone of the network system. It happens.

Another indication of shifting power is the number of journalists who are abandoning the networks for local stations. It's a long parade of faces and names: Sylvia Chase to San Francisco, Bill Redeker to Los Angeles, Emery King to Detroit, Bill Kurtis to Chicago, Lee McCarthy to Boston, and Morton Dean to New York.

Yet when local news directors and network executives are asked how well they get along together, most of them will say, "Swell." The Axis said the same thing about the Allies after surrendering in World War II.

THE STATIONS' POINT OF VIEW

There isn't actually much left for the formerly downtrodden stations to complain about. Reagan Ramsey of Portland's KGW says, "Our relationship with NBC is fabulous. We work together on coverage outside of market areas. When we go overseas, the network gets to the bureaus and cooperation is very good. The bureaus provide any footage we need, even dub it off for us." Not quite a two-way street.

As for scooping the networks and pouncing on non-local breaking stories on their newscasts, that's the way it is today, maintains the Portland news director. "Networks have to understand, in a big breaking world news story, we're on before they are, so we're going to grab the story and run it." The networks may end up with egg on their face by running the same story on their shows, admits the King Company executive, so stations should at least make it look different. According to Ramsey, however, many stations don't even bother doing that. They simply cannibalize what they want from "Nightly News."

KGW's approach is becoming that of stations throughout the United States. They utilize all available tools, including the material fed down the line by the streamlined syndication services. As a result, local producers can build a far better story than the networks can. Because network news has a two-or three-hour time difference to cope with, more affiliates are moving in on the live breaking news area.

So far, news directors like Ramsey avoid competition with their networks on the world news beat in the area of breaking stories. The local news executives admit they don't have the reportorial expertise or logistics to cover major stories overseas. A vivid example of stations trying to compete anyway involved one of the major international stories of the decade. The networks reacted swiftly and professionally to the November 1989 disintegration of the Berlin Wall, but many local newsrooms in the United States saw it as an opportunity to impress viewers. One reporter, from KABC-TV in Los Angeles, ended up delivering his live reports about the Wall from *Frankfurt*. It is not coincidental that November is a crucial ratings period, when surveyed viewers rank the popularity of the newscasts.

For many news directors, the networks' syndication service isn't enough. Although the networks are trying to provide more national and international material to the local stations, Jeff Rosser accuses them of continuing to hoard video. He refers in particular to material that could be made available to him, but that doesn't appear until the evening network newscast—well after his local newscast.

Rosser, however, doesn't think local stations are replacing the networks. He, too, acknowledges that affiliates don't have staffers who understand national and international events or can "make the kind of editorial decisions that need to be made, to ask the questions that need to be asked."

KING-TV's Donald Varyu also describes his relations with the network as "very good." The Seattle station's relationship with NBC's West Coast bureau is "absolutely terrific" and KING never gets flak when it gets into international news content.

Varyu agrees that stations are tending to cover the breaking news in their areas while networks continue to pursue the in-depth national and international stories. "I don't see a lot of stations out there trying to do the Pentagon scandal because nobody is going to try to do that on the local level."

Neither Varyu, whose station is affiliated with NBC, nor KGO's Harry Fuller, who works for an ABC affiliate, hears network death rattles. Fuller, in San Francisco, feels local stations can provide the breaking national and international news, thanks to services such as CNN. He also agrees they can't come up with the sophisticated reportage and facilities available to the networks.

However, many affiliates insist the lavish feed services and Ku-band truck joint ventures aren't enough. There is an increasing demand for the network to provide pieces from overseas bureaus done especially for the stations. They want seasoned network correspondents with specific expertise made more accessible to local stations.

If these demands are met, the networks are likely to find themselves changing roles. Many experts think such a change is inevitable. The network would become more of a wire service, feeding the needs and whims of the local stations—kind of an Associated Press of TV.

Since practically all of their demands are met, station news executives are very positive in their description of local–network news relationships. The situation will probably remain that way as long as the affiliates continue getting what they want.

THE NETWORKS' POINT OF VIEW

The network brass have other reasons—besides wanting to keep the affiliates and besides the simple inevitability of it all—for going along with the current trend. They need the stations' help to **tease** and promote the big show. NBC's Jo Moring for one claims that providing the stations with network newscast material helps "Nightly News": "The network recognizes that a tease is better than no tease at all. If you have some footage that a station can use to cover the story, it also helps lead the flow to 'Nightly News.'" Moring does not think the new affiliate services are watering down the network product. Network news, she feels, is still trying to present a very tidy half hour that gives you "a whole day's news plus introspection, reflection, and commentary."

ABC's Paul Dolan is also realistic about the stations' new power. "Whatever we think is beside the point, because stations make their own decision and they have the resources." To Dolan, the local stations' involvement in international stories is a reality, and he tends toward helping them do a better job.

Network correspondent Ed Bradley takes it further. CBS's "60 Minutes" reporter told the RTNDA session that he is encouraged by local television around the country. He praised their extended reports and documentaries, as well as the fact that "more stations [are] sending reporters to cover foreign stories as well as the political conventions and campaigns." Bradley sees the local broadcasts as a vehicle for improving election coverage. While the networks are faced with coverage tailored to TV by the politicians, local news has more freedom. The stations are innovators, Bradley feels. He wants them to concentrate on those stories that occur outside the community but have an impact on the lives of people in their market.

Call it wishful thinking, but Robert Mulholland, former president of NBC news, thinks the two forces are successfully staying out of each other's way and living happily ever after. Mulholland, who watches TV news whenever he's on the road, clearly disagrees with the assumption that local stations should or do cover all the news before the networks. The executive, who has lived in both local and network worlds, claims stations don't have enough room in their newscasts to do national and international news. "Why should they and squeeze out the local news [in that] national and international is going to come up afterward on the network?"

LESS DIPLOMATIC VOICES

Some news directors, including Larry Price of KSTP in Minneapolis, are somewhat more candid and less diplomatic about the changing role of the networks. Price is among the many who feed materials from the network, as well as from his Ku-band truck, and an independent service, Conus, into his newscasts. He also aligns himself with those local news executives who declare "the years of network rule and network arrogance over." However, the KSTP news director thinks some affiliates have taken on "a bit of the same arrogance." Price predicts networks will become more of a feed service somewhere down the line.

The drive toward self-rule is reflected in other areas as well. Gary Mehalik says the networks no longer have a chokehold, for example, on manpower and cameras. "Now the affiliates, or local stations, have more cameras out there, know the turf, know who to talk to and sometimes bring back better stories." As far as Miami's WTVJ is concerned, it means more clout for the stations and "the realization that we are the network."

Not that WTVJ and the network don't cooperate. According to the assignment editor, they help each other with crews. That means a network crew can end up covering a local-local story while a WTVJ team could be on a national one.

Seattle's James L. Boyer is among the news directors who believe it is possible network news is dying, but says it's not his fault. To Boyer, it is really a matter of availability of material. While KOMO is doing local news for a local audience, the station sometimes gets into traditional network areas. To Boyer, this occurs not because his staffers can do it better, but because they control the delivery systems.

Even the have-not stations, like Helena, Montana's KTVH, see the affiliate's role as shooting local breaking news stories for the network. To news director Ian Marquand, it makes sense for the network to rely on the affiliates for local coverage. That's a far cry from years back, when New York would send its own crews, reporters, and producers to cover major breakers in the station's province.

And in Salt Lake City, J. Spencer Kinard says his CBS affiliate is still dependent on the network for certain stories but, "if I have to, I can provide my viewers with most of the information they need and the pictures they need." KSL-TV has a bureau in Washington, D.C., and contacts in Europe to cover stories there. Kinard occasionally reflects back to the time when the network would cover anything that was important in the world. But times have changed. Today, local stations are not "just running around with a **Bell and Howell** (film camera) chasing a fire truck. We're covering the important stories as well and we have the capability and the sophistication to do it." Nor will Kinard take the blame for killing the network. Actually, he doesn't think the webs will die, although he foresees them in a more diminished role.

Most news executives agree that certain stories still have to come from the network. Local stations simply don't have access to Shanghai, Moscow, or other difficult beats. For example, Jerome Nachman, former New York and Washington, D.C., TV executive, who didn't recognize international borders during his coverage days, concedes that he can't handle certain stories, "even as a wealthy and fat 'O and O,' [network owned and operated station] because, when I have a story in the Near East that requires a peg, perhaps in Amman, another one in Tehran, another one in Tel Aviv, Jerusalem, and maybe London and Washington, I can't do that."

Sensible local news executives acknowledge they can't compete head-on with the networks. Jim Snyder says he and his four Post-Newsweek stations "kill ourselves to cooperate with the network. We want the network to be healthy." So why don't news departments stick to what they can do best, covering local news? The news vice president thinks the reason is promotional advantage over local competitors, not over the network. But as far as Snyder is concerned, the audience relates best to the station that is doing the best local news. He feels that provides a local station much better grist for promotion than bragging about being first at the Geneva summit.

RTNDA's Ernie Schultz predicts *rapprochement* between the networks and local stations. While the balance of power will probably continue shifting toward the affiliates, Schultz thinks it's "crazy" not to use network resources. He agrees networks are not going to vanish, "not if they find out what their strengths are and emphasize those strengths and not try to recreate the golden age of network television news, whatever that was."

OVERSEAS, TOGETHERNESS

Foreign networks and local stations, for the most part, are one happy family—so far. But those grand old government owned and controlled dynasties are breaking up with the increasing trend toward privatizing. Just as significantly, overseas broadcasters consider the United States way ahead of the world in commercial TV news and emulate accordingly.

Countries that have a network channel dedicated solely to local and regional news report little or no conflict. In addition, the non-government entrepreneurs—in Italy and France, for example—so far haven't figured out a way to make news profitable.

Japan, particularly, is proud of its network–affiliate relations. The government's NHK and the commercial channels consistently report no conflict between the networks and the affiliates. There is friendly communication and a productive exchange of footage, people, equipment, and facilities.

In Great Britain, however, there are potential problems. Unlike with the BBC, where the network comes first, ITN and its regional operations can be competitive. Barrie Sales says his Thames crews occasionally shoot stories that ITN wants. The regional news bureau will shoot and air certain stories first—more than four hours earlier—whether the network wants the coverage or not. However, on other occasions, the two cooperate, sharing crews and footage at major breaking stories in the London area.

Despite trends in the United States, there is little movement in Great Britain toward regional networking. If a story affecting London breaks in another city, which happens occasionally, an exchange will be worked out. However, Sales is adamant about concentrating on his own beat and notes the government's role in seeing that he does. "We all stick to our own patch. We virtually have to because of our franchise, as far as local news is concerned. We don't poke in other people's area."

It's equally peaceful between local and network operations for BBC in Birmingham. If there are any problems, they are resolved on a daily basis, says Rick Thompson. "We keep the attitude that national comes first, that the primary call in our services will be the national news." As a result, Thompson's operation often beats itself by allowing the BBC network to air its material first. However, the local bureau tries to work out a different angle and style for those watching both programs.

PROJECTION

It's eerie, listening to attitudes that were once commonplace in U.S. broadcast journalism and knowing that someday the news bosses in these countries will probably be singing a different tune. Chances are they'll be the same notes that resound in the news corridors of America today. High tech. Ku-band. Go anywhere. Cover anything. Show the flag. Diminishing network.

As for our broadcast journalism, it faces an era of change, adaptation, a certain amount of attrition, a generous barrage of challenge, and the opportunity for an even more innovative, viewer-friendly TV news menu. As profit continues to be an overwhelming motive, it will generate much of what happens to local and network news.

For the next decade, local stations will expand even farther nationally and overseas, particularly in the pursuit of a story with a local angle. Such expansion may be reasonable, such as when going after a story of interest to the local audience, or it may not, such as when promoting an anchor or reporter.

At the same time, the budget-conscious networks will probably continue cutting back on overseas bureaus. In addition, network correspondent stars will continue to fall as the brass permits anchors to increasingly diminish the role of the field reporter. For example, the networks rushed their anchors to the dramatic turmoil in Tian An Men Square in Beijing in 1989. The coverage exemplified the best of network professionalism, drawing praise from media critics and viewers across the country. However, continuing efforts to enhance the anchors' profiles have a negative impact on the status of correspondents. These efforts could lead to an even bigger exodus of star reporters from network. Of course, the positive aspect if this is that such a migration is certain to improve the quality of affiliate coverage on any beat.

Also on the plus side, viewers will get far more complete newscasts on their local stations, thanks to the availability of program elements. Such elements include the local station's own field coverage as well as the myriad services and feeds that already cover the world. But as newscasts take on the one-stop shopping characteristics of newspapers, local-local news will temporarily suffer.

At some point, the cost of covering the country and the world will become prohibitive and local station managers will tighten the leash. Hopefully, the stations will take the very same technology—SNG, SNV, computer graphics—that lures them away and use it to cover local news even more incisively than before.

However, the stations' breakneck race for supremacy will take a toll elsewhere. Expenditures on coverage and technology will grow out of control, forcing the weaker news operations in many cities to fold and their stations to invest in other programming potential.

Nor will the network news operations necessarily be among the fittest to survive, at least not in their present role. Profits simply cannot be divided fairly among the local stations, cable, and the networks. There are several paths open to the networks, ranging from self-destruction to galvanization. *Life* magazine kissed off the networks in its February 1989 forecast for the year 2000, sounding the death knell for them as well as for "manual typewriters and disposable diapers." The *Life* obit is a simple one: "Competition from cable and entertainment systems catering to highly individual tastes may deliver a TKO to television's Big Three." Reuven Frank, former president of NBC news, is only slightly less pessimistic. He told *Playboy* magazine at least one network will drop "straight news presentations in the early 1990s."

On a more positive note, network brass could consider streamlining their news product with alternative programming. Although local stations are rapidly cornering the breaking news market, they still respect network's ability to provide in-depth coverage. Instead of continuing the wilting daily half-hour newscast, the nets could instead deliver a format out of reach for the affiliates. For example, the networks could go for the one-issue type of program, such as "Nightline," or the news and depth combo, such as "This Week With David Brinkley" and the "MacNeil-Lehrer NewsHour." With a change of format, the networks could also consider abandoning the early evening time period for the late evening, network-controlled block. As it is, few stations schedule the network newscast for the traditional 7 P.M. slot. They move it to whatever spot in the afternoon block is convenient to their shows.

The most likely future for network news is for it to continue in its present direction of serving the affiliates. As a result, roles will completely flip-flop, with the networks emerging as an appendage of the stations. In effect, they will become **wire services**, like the Associated Press, feeding visuals and reporter pieces to their stations. Early signs of this trend already exist. NBC has bought a large percentage of Visnews, the British overseas news operation that serves 400 broadcasting companies in 84 countries. The majority owner of Visnews is a wire service, the Reuters news agency. In buying into Visnews, NBC catches up to ABC, which has 45 percent of a similar operation, Worldwide Television News, also based in London.

Reuven Frank, whose 38 years at NBC News included two tours as president, was quoted in the August 2, 1988, *Los Angeles Times* as predicting the wire service role as the networks' destiny, with "people at the [local] stations" doing the rest of it. Other forecasts vary, and, although few are as grim as that of *Life*, there is a feeling that, in network news, only the fittest will survive. That means at least one of the big three is in danger.

Another alternative for the networks is to join, and not fight, cable.

NBC, for one, has already taken the initiative in that direction with its $60 million investment in CNBC, or Consumer News and Business Channel. The network's 24-hour-a-day cable operation, starting off with 13 million homes, was launched in spring 1989. Network executives deny CNBC is phase one of a master plan challenging FNN, CNN, and the proliferation of other competitive cable services that specialize in such areas as sports and weather.

It is a safe assumption that by the year 2000 the news hierarchies of both local and network operations, as well as their relationships with each other, will be dramatically different. Of all the possible scenarios, the emergence of the network as the complete service to affiliates, which in turn will become providers of the complete news budget, is probably the most likely. It might in the long run also be the least harmful. That doesn't mean all the network and all the local news operations will survive. Those engaging in a mature approach to the concept of joint venture will make it—as will the fittest and most profitable. With luck, the survivors will also be seriously dedicated to the profession of journalism.

4

---■---

A FIFTH AND SQUEAKY WHEEL

The TV News Consultant

"The consultant beckoned with a jeweled finger; he invoked the sweet, mysterious patois of behaviorist psychology and then preached the ringing, pure gospel of profit swift and certain. Done, said the managers." Ron Powers, The Newscasters.

If television critic Ron Powers's exposé, *The Newscasters*, were required reading for graduating seniors, the real world would be less of a shock to entry-level broadcast journalists. There might also be fewer broadcast journalists. Newcomers, warned that they would, in effect, be reporting to an additional, often invisible tier of bosses, could very well have second thoughts about the profession.

Consultants invaded television in strength in the early 1970s, when TV news was becoming lucrative. Managers hired them to determine which people the audience preferred to have delivering the news, exactly what news they would deliver, and how they should mold the product accordingly. To the general managers, most of whom at the time had moved up the ladder from the sales division, it made supply and demand sense. The tradeoff, however, would shake the very journalistic foundation of TV news. From here on, producers would be forced to give the viewers what they wanted, not what they should have.

The managers had their astrologer and, reports Pulitzer Prize winner Powers, his sign was the dollar. "Armed with questionnaires, with the rudiments of Gestalt—and, in some cases, with electrodes—the consultant set about to spy on the viewing audiences for the managers, to pry into behavior patterns, to pilfer the unconscious if necessary; but above all, to find out which stimuli (faces, voices, colors, names, jokes, bedtime stories, charades, or, God help us, *ideas*) would serve as the best bait to lure the viewer before the Client's Channel."

CONSULTANTS IN THE 1970s

For us, the news management at KABC-TV in Los Angeles in the early 1970s, it meant a double whammy, because the ABC-owned station bought the insurance of the *two* leading consulting companies, McHugh & Hoffman Inc. of McLean, Virginia, and Frank N. Magid Associates of Marion, Iowa.

This meant ongoing written critiques about such things as anchor delivery and appearance; reporter delivery and appearance; newscast item count; choice of stories; **stacking** of the newscast; quality of shooting, editing, and writing; set design; effectiveness of teases and headlines; and the clothing and makeup of on-the-air personalities. As Channel 7 was *"Eyewitness News,"* an analysis of whether anchors and reporters were truly "involved and really touching, feeling, and smelling" the story, as one executive frequently put it, was a running theme for all critiques.

Separate written reports from both consulting companies arrived monthly based on TV cassettes of newscasts sent to them or on their viewing the newscasts live during overnight stays in Los Angeles.

Naturally, our product would be compared with that of competitors, and it would be carefully noted if ours lagged behind theirs in the number of stories in the newscast, the item count, whether our show got into videotaped or live reports as fast, or whether we missed a

particularly juicy crime break. The report card would also be peppered with grades on our how-to-cope features, the disease of the
week, folksy weathermen, or jumping jack sportscasters, depending
on what trends happened to be plucked from markets elsewhere in the
country at that particular time.

Then, and for years afterward, KABC's consultants never set foot
in the focal point of all the fuss, the newsroom itself. Meetings,
attended only by the consultants, general manager, news director,
executive producer, and assistant news director, always took place in
the GM's office. We were shown the written reports, but not permitted
to keep them, and they were never directly referred to as the sugges-
tions became orders passed on to the lower echelon of producers,
writers, and on-air talent. Nor was open discussion encouraged at
critique sessions. Newsroom management was given the opportunity
to react to these outsiders who had no journalism portfolios, but it was
understood that the proper response was acquiescence. The situation
was somewhat like the highway patrolman asking for your opinion
while making out the speeding ticket.

The same fait accompli despondency permeated the consultant-
inspired search for a new anchor team. All the stops came out for that
move, although often the process simply confirmed a management
decision already made. Such was the drive that married a stunning,
young, exuberant blonde beauty with a market-experienced, white-
haired father image. They were obviously a winning team. We knew
it. They knew it. The consultants knew it. But for political reasons,
more likely related to anchor incumbents and contenders, KABC-
sponsored research had to *show* they were a winning team before the
lovable couple could settle onto their thrones. Pollsters hit the streets
flashing photos of various TV personalities at unsuspecting pedestri-
ans and asking for recognition ratings. Next, videotapes of various
anchors were edited together in different combinations for test viewing. Anchor A doing a newscast would be intercut with Anchor B
doing an entirely different one, just to check out "the chemistry." That
they were entirely different newscasts aired on different days forced
together through the magic of videotape editing did not faze anyone.
The tapes were then shown to sample audiences at special screen-
ings—like market testing a new sitcom. The viewer sat in a specially
rigged seat armed with buttons that registered approval or disap-
proval of the person appearing on camera. In addition, viewers were
clustered into small focus group meetings, where discussion leaders
coaxed them into opinions about various anchor possibilities. Executives were permitted to listen and watch through one-way mirrors or
to wait for the transcribed audio tapes.

The coronation quickly followed and Los Angeles viewers anointed the couple with years of profitable rating points.

THE SKIN TEST

Other consultants in other locales used even more shocking methods to determine what the public wanted.   One device that particularly enraged media critics was the one that became known as the skin test. The sample audience's actual chemical and emotional reactions to videotapes of various stories and news personalities were determined directly, through electrodes attached to the viewers' fingers. The human guinea pigs' emotional reaction to the program was then re-corded in graph form on a fancy machine in the next room.

The flurry of bad press that this created did little to discourage the flourishing news consultants as they and station managers cried all the way to the bank. Word that the formula worked traveled fast through the maze of network and independent stations. As Barbara Matusow points out in *The Evening Stars*, within a few years almost every station in the country either had retained its own consultant "or was copying the stations that had consultants."   The result was a sameness in format, content, look, and delivery of news from coast to coast. Ma-tusow's list of features that stations across the country had in common includes:

- a faster-paced newscast;
- more emphasis on local news;
- more sensationalism;
- greater emotional content;
- a consumer ombudsman;
- advice on coping;
- reporter involvement;
- more humor;
- longer newscasts;
- a friendly anchor team.

CONSULTANTS IN THE 1980s

Today, a local news department without consultants—and that result-ing homogenized look they create—is still as rare as it was in the 1970s and late 1980s. The main difference now, more than a decade later, is that there are many more consulting companies, and the old-timers, like Magid and McHugh & Hoffman, no longer hide in the general manager's office.

Many of the newer consulting companies, several of which are one-person operations, specialize and are even referred to by news directors as consulting "boutiques" because they work with a small clientele. That's the way KOMO's James L. Boyer in Seattle describes his consultant, Joe Barnes of San Francisco. Another popular consulting company is Audience Research and Development of Dallas, specialists in local-local news coverage, according to client WCCO in the Twin Cities.

Most news departments surveyed display their consultants proudly. In KSL's Salt Lake City newsroom, for example, the Magid experts hold training seminars for producers, writers, reporters, and photographers. The sessions get as specialized as how to write a tease and how to write in the active voice.

The roll-up-the-sleeves-and-work-with-the-troops-in-the-newsroom concept is relatively new for Magid. But at least one other company, which is considerably smaller than Magid but has been around just as long, has excelled in the practice from the beginning. The Virgil Mitchell Group in North Hollywood, California, is unique in other ways as well. Mitchell, a former news director, and his associates have long, admirable records in broadcast journalism. When the consultant craze started, the bigger companies were staffed primarily with researchers rather than seasoned journalists. The Mitchell Group, on the other hand, has taught the old-fashioned journalism basics of news judgment, writing, producing, reporting, and anchoring—and done so right in the client's newsroom—all along.

Mitchell abhors another popular tool of the larger consultants, the minicourse in instant newsmanship. The device, which KABC-TV started using back in the mid-1970s, is tailored for cosmetically attractive reporters and anchors who lack talent in writing, reporting, or anchoring. It's bad enough that nonjournalists are hired for these broadcast jobs in the first place, but the real insult to the profession is the attitude that they can learn it all in one overnight stay in Magid's Iowa headquarters.

The Magid crash course is also used to transform anchors and reporters into more commercially acceptable images. Veteran WNBC-TV journalist Bob Teague, in *Live and Off-Color: News Biz*, tells the story of a WABC-TV reporter in New York. The woman, a "dark-haired, copper-toned Puerto Rican beauty," was flown to Iowa for one day of lessons. Her instructor was "a kid about 22 years old, a guy who had never worked in television." The reporter told Teague the instructor took her into a news studio, gave her some copy to read, then critiqued her performance. The reporter said she was told "to start making love to the microphone." When the reporter told her instructor that was dumb, he showed her videotapes of Jane Pauley of *Today* "and a few other sweet middle-American girls who were doing it right." The next time around she cornballed it, "the old goodness gracious Howdy Doody bit." The result: "The kid bought it, gave me a passing grade. I flew back to New York and tried to forget the whole thing." The reporter was Gloria Rojas, a 10-year WABC-TV news veteran at the time.

THE STATIONS

Not all news directors submit completely to this rape of journalistic responsibility by upper management. KGO's Harry Fuller, for example, is under pressure from brass and McHugh & Hoffman to use his expensive, promotable electronic toys to go live more often. Despite insistence from the consultants "that live TV is more interesting than tape," the San Francisco news director persists in using his microwave equipment judiciously.

However, McHugh & Hoffman does help Fuller combat what he refers to as a newsroom full of "yuppie journalists, people who want to put the *New York Times* on the air." Thus, Fuller uses the consultants to tell his staffers, "It's not a sin to be interesting. It's not a sin to do interesting television that draws people into the programs."

In Portland, Matt Shelley of KGW-TV says that consultants there don't completely dominate the operation, either. Shelley, who uses Magid, claims KGW-TV doesn't put much emphasis on the consultant. The Iowa experts come up with research about other markets, "but don't come in to tell us what to do." As for Magid criticisms of his shows, "some we listen to, some we don't."

Other stations are completely off-limits to consultants. San Francisco's KRON-TV doesn't employ them, and KING-TV is proud of having unloaded its company five years ago. Donald Varyu blames weaknesses in the coverage of his Seattle competitors on their use of consultants. Varyu says ABC affiliate KOMO and CBS's KIRO follow consulting company formulas that stress the emotional aspect of stories. As a result, they tend to emphasize stories that are only "viscerally important," saturating newscasts with sensational murder cases or other tragedies.

While Jim Snyder doesn't hire outside help for news departments at the Post-Newsweek stations, he does have a benign attitude toward the concept. He thinks consultants are beneficial for "a medium or small market station, where the manager may be an ex-salesman and the news director may have been on the job a month and a half." However, Snyder has no tolerance for local managements that surrender all journalistic rights to consultants.

The RTNDA's Ernie Schultz diplomatically defends the role of the consultant. He feels that "in every case," when they've come into a market, "they've done something for the station." As for the tendency to create a sameness in the news product from station to station, Schultz agrees, but points to the increasing sameness in newspapers as well.

OVERSEAS

The consultant is one U.S. product that, so far, seems not to have traveled beyond our own shores. There was no evidence of consulting companies in the countries surveyed for *Changing Channels*. Actually, many of the news executives, particularly those in Japan, were adamantly negative when asked if they hired or were planning to hire consultants.

PROJECTION

Consultants, in every case, do "something for the station." Unfortunately, the "something" is not always positive for journalism. Nevertheless, the practice will continue flourishing. With the big consulting companies getting bigger, and the number of small firms increasing, the danger mushrooms accordingly. It could mean an even wider spread of formula newscasting and its accompaniment: journalism by the numbers. As consulting companies make deeper inroads into newsrooms, network as well as local, news management abdicates its precious responsibility to monitor its own product, teach its own novices, guide its own operations, and discipline its own errant staffers. It also abdicates its responsibility for long- and short-range news decisions. It is far too easy for networks and stations to make an unpopular or questionable decision and have it backed up with consultant "research." When a station has the blessing of the outside "expert," it is also too easy to legitimize bad journalism in the name of profits.

A case may be made for getting strictly professional help from consultants of integrity for small news departments with inadequate journalism talent. However, a stronger case can be made for taking a closer look at overall hiring and consulting practices. It could make more sense to take the money being paid out to consultants and use it to hire more capable journalists.

5

---□---

THE BUCK STARTS HERE

The Commercialization
of TV News

There is a dangerous correlation between profitable news and its increasingly popular content. While the gold pays bills, the ratings race and greed could destroy the goose, journalism.

In the 1960s, the networks dominated TV news, but it was not a profitable operation. Media critic Jeff Greenfield credits the networks' profitable 1970s to several factors: the entry of ABC as a serious news contender, the commercial success of CBS's "60 Minutes," and the one-minute prime time "News Updates." In 1970, the three network news divisions were losing approximately $20 million a year, but nine years later, in 1979, they earned a profit of about $760 million. In the 1980s, they were back in the red for a variety of reasons, including the high cost of covering major breaking stories, such as sensational plane hijackings; skyrocketing talent and equipment costs; competition from cable and local stations; and economy-minded new owners.

Greenfield, in the Gannett Center *Journal's* special Spring 1987 edition on "The Business of News," reports a far more volcanic eruption for local TV news. From 1965 to 1972, local TV advertising grew 150 percent, while that for the networks went up 50 percent. At the same time, high tech was replacing film with videotape and bringing the size and cost of camera and editing equipment down to a level that was reachable for the stations. The money-making clincher, however, was the popularization of news with the **happy talk**, Eyewitness News style of packaging.

The more entertaining news became, the more money it made for the stations. By the 1980s, a majority could credit news with more than half their profits. In 1988, more than two-thirds of 459 TV news directors polled by the RTNDA reported that their stations made money on news. Vernon A. Stone's survey showed that nearly all network affiliates in the 25 largest markets were profit centers.

A study by Wilkovsky Gruen Associations for Blair Television, "TV 1995: The Future of the Television Industry," has predicted that advertising on TV stations will increase by $16 to $30 billion by 1995. The same study had bad news for the networks, forecasting that the big three's share of total TV advertising will fall to 30 percent by 1995, a drop of 5 percent. Cable will continue flourishing, its advertising mushrooming threefold to an impressive $3.4 billion in 1995.

IMPACT OF BUYING AND SELLING

Unlike the 1960s, when broadcast journalists were concerned about news copy and not profit-and-loss statements, the pressure is on both network and local operations. The buying and selling of networks and stations, practices considerably loosened by government policy, are making life even more difficult. The situation is not unlike that of a family buying real estate without adequate financing: The underfinanced family squeaks through escrow and gets the house, but not the needed furniture, paint job, or landscaping. For TV stations, this translates into a heavier emphasis on budget cutting. Too many new owners go into hock buying into TV on the expectation of making money. Then they see looming in front of them a high profile of fat, the flourishing news division.

Take the case of WZZM-TV in Grand Rapids, Michigan, a sad tale unfolded in the December 1988 *Washington Journalism Review* by investigative reporter Mark Lagerkvist. WZZM Channel 13 was number one in news ratings and awards until Price Communications Corporation bought the station in 1986. Lagerkvist reports that the entire $62 million purchase of the station was a leveraged buyout, "almost totally financed with borrowed money." Price paid interest on the debt by slashing budgets, personnel, salaries, and employee benefits.

Lagerkvist himself left his job as an investigative reporter because of the mood created by the corporation's Robert Price. The boss simply killed investigative reporting, saying it is "the type of journalism that I don't want in my company." Also killed were the station's two-person public affairs department, the twice-a-week editorials, and the early morning newscast. Within two years, about half of WZZM's on-air personnel had quit or were fired. And within the same time period, WZZM was replaced by rival WOTV as market news leader.

WINNING AT ANY COST

Football coach Vince Lombardi started something by preaching that winning is everything. Too many station executives, taking full advantage of their newly discovered profit centers, apply the same philosophy to TV news, thus creating incredible pressure on news personnel. A case in point is WTSP-TV in St. Petersburg, Florida, where two news executives were fired in a case involving tapping into the newsroom computers of a competing station. The assistant news director was arrested and charged with repeatedly breaking into the newsroom computer of rival WTVT in Tampa for the purpose of raiding its confidential news files. Because of the arrest, during the February ratings sweeps, the assistant news director as well as his boss, the news director, were let go.

Elsewhere throughout the country, crimes against the profession of broadcast journalism in the name of profits are just as devastating.

When we news executives were called into general manager John Severino's office at Channel 7 in Los Angeles back in the 1970s, it was either to have an after-the-show drink, play liar's poker, be critiqued by the two consultant companies, or be browbeaten into enticing more viewers. The simple equation was sensationalism = viewers = ratings = profits. The unspoken corollary was that it didn't matter how you got to the end result: bucks. If anybody had the courage to suggest that this approach was not conducive to good journalism, the response was, "Journalism has nothing to do with this." It was a philosophy that became obviously apparent in the behind-the-scenes activity and the on-the-air result.

It was not unusual, for example, for the general manager to hold in his hand the summaries of the lead stories from the previous evening's newscasts produced by the other four ABC-owned stations across the country. "Look at this," he would say. "Chicago opened with three southside murders, San Francisco with the finding of human bones," and so on. "Sev" would go on to show us how each of the stations showed their enterprise by digging up the sleaziest, bloodiest crime story and hooking the audience by leading with it and promoting the hell out of it in the teases.

THE EXPLOITATIVE MINI-DOC

The ratings sweeps, that four-times-a-year frenzy to destroy the competition and come out first in "the **book**," or **ratings book**, were even more blatantly anti-journalism crusades. Executives from outside the news department suddenly emerged and became an integral force in the decision-making process. The biggest fuss was over the news executives' proposed list of **mini-docs**, those serialized versions of documentaries aired over several newscasts to garner ratings. Meetings to choose the subjects brought in the sales, promotion, and research departments, as well as the consultants. They all had to like the proposals from their various points of view before the general manager, not the news director, gave them the go-ahead.

Nor did their influences stop there. A proposal by the author entitled "The Million Dollar Martini" had all the basic ingredients: a solid mini-doc that was informative as well as exciting. The basis was a gruesome accident involving actor James Stacy. The actor had been riding his motorcycle down a Los Angeles canyon road when he was hit by a drunk driver. A passenger on his motorcycle was killed and Stacy lost two limbs, an arm and a leg. The accident resulted in a clamor and serious consideration to hold liquor establishments and hosts of private parties legally and financially responsible for drunk driving accidents caused by patrons and guests who were served too many drinks.

It was a serious documentary that deemphasized the sensational aspects of the accident and concentrated on how the proposed law would affect social and business lives. Stacy's involvement was respectfully limited to a short on-camera statement. However, the ad promoting the mini-doc included a photo of the actor and the headline, "The Million Dollar Martini. *It could cost you an arm and a leg!*" Efforts to change the ad were ignored. It was the era when station managers throughout the country had already realized there was money in local news. Money and sleaze.

It would be nice to say those horror stories were just another phase, that the TV industry has since corrected itself in midflight and lived happily ever after. But tune into the news any February, May, July, or November, and you are likely to get a series on every conceivable aspect of sex, man to woman, man to man, woman to woman, employer to employee, brother to sister, parent to child, rapist to victim, old person to old person, married to single, black to white, race to race, and so on ad nauseam. Or 20 different ways for bald men to grow hair. Animals? That covers the gamut from what's the best pussy cat to how to get your dog's head shrunk. Or diets. Or nude beaches. Or prostitution. Or the best swim suits. Or what's in a hamburger, ice cream, peanut butter—or just choose from column A or B. Or the complete visual diary of the birth of the anchorman's or woman's baby or babies. Or any subject that titillates, warms the heart, promotes, or simply looks seductive in print ad and on-the-air tease form.

The same Channel 7, some 15 years later, is no less sensational or exploitative. It lifted eyebrows with its "Quakebusters" pieces during the May 1989 sweeps. The idea was to simulate live coverage of an earthquake—the big one—on all its afternoon newscasts. The "tragedy" and its reportage was *acted out* by the Eyewitness News team.

As Ron Powers maintains, the biggest heist has nothing to do with banks, but with the newscast itself. "The salesmen took it. They took it away from the journalists, slowly, patiently, gradually, and with such finesse that nobody noticed until it was too late." Responses to charges of sellout and commercialization from news executives vary in intensity and honesty. Many admit their business is increasingly about money. Others are upset because critics don't recognize the good things news departments do the rest of the time. And still others try to cool off the heat by waving "sex and sleaze" lingerie and mudwrestling ads from *newspapers*.

Bill Sternoff, who spent 20 years in national and local TV news, accused news directors, at their 1988 convention, of selling out "your staff and the news department for a promotional scheme of a supermarket." Sternoff has clairvoyant reportorial abilities. Within a year of his observation, KNBC-TV in Los Angeles made a deal with the May Company department stores. Channel 4's news anchors appear in the May Company's print ads for TV sets. The chain gets commercial time in return. What also gets to Sternoff is the practice of promoting shows within the body of the news program. Both local and network newscasts tend more and more to work in features related to non-news programs appearing on their air. Add to that still another unfortunate tendency, that of news shows reporting only awards won by them, and journalistic credibility stretches to the breaking point.

It is possible that ratings sweeps promotions, including the blatant titillating approach of pushing mini-docs, could backfire on the profit seekers. Post-Newsweek's Jim Snyder for one believes the audience is too savvy. "They know when you promote something and don't deliver anything. They know that you screwed up."

IMPACT ON POLITICAL COVERAGE

Author Mark Hertsgaard includes the commercialization of local and network TV news as among the causes of the media's failings in covering the 1988 presidential campaign. George Bush and Michael Dukakis, Hertsgaard wrote in *On Bended Knee: The Press and the Reagan Presidency*, controlled the message by staging photo-opportunity events, keeping reporters at a distance, and avoiding being drawn into meaningful give-and-take. "Ratings prevail over responsibility, news is treated as a commodity to be sold rather than an educational trust to be fulfilled, and fundamental questions about the nation's direction are neglected in favor of the six-second sound bite."

During campaigns, too many candidates take advantage of the local stations' thirst for footage to fill newscasts. They favor local exposure over that of the network because, regardless of how meaningless, the events are thoroughly covered. More importantly, the affiliate's drive to show the flag and show off an anchor who is often inexpert in politics leads to far less dangerous questioning than that of a hardball player like ABC's Sam Donaldson.

BRINGING THEM INTO THE TENT

Among the news operations that have matured but still recognize the need for audience and ratings points is ABC-owned KGO-TV. Back in the early 1970s Channel 7 in San Francisco reached its lowest point with a highly teased film story about a severed penis found on the railroad tracks. Today, Harry Fuller insists, TV news has veered away from the cheap sensationalism of that era. "In those days there was a certain portion of the audience that watched because we were the nastiest, raunchiest, most unbelievable thing on television." That audience, the news director maintains, now can rent the porno or violent movies instead. "We're not titillation any more."

However, critics who reflect back on the networks' golden era when producers determined what was best for viewers would have a field day with some aspects of the Fuller philosophy. It is typical, they would say, of an attitude that jealously defends present trends and has no use for the good old days. But, according to Fuller, the 1960s were the "holier than thou, shut up and sit down and listen to what I got to tell you kid because otherwise you're an ignorant fool" attitude. This, he points out, has been replaced on the local level with a more mature approach to informing the public. His emphasis is on how events affect the viewer's life, pocketbook, or children's education.

Fuller's problem is that KGO's staffers, most of whom were born after World War II, believe in "proper journalism." And "proper journalism," as far as he is concerned, is the "MacNeil-Lehrer NewsHour," which the news director describes as "boring as hell and the average TV viewer is not going to watch it even if you hold a gun to his head." So, as far as which force, the broadcaster or the viewer, will determine what goes on the air, Fuller maintains, "You can't be journalistic above all else. You have to be television."

At least one network executive partially agrees with him. NBC's Jo Moring is realistic about the commercial pressure to make good. She compares the highly competitive news directors to the religious revivalists who have "to bring them into the tent." In other words, in today's hard business world, the best story in the world doesn't mean a thing if nobody sees it. "If it takes having a pretty anchor woman, if it takes having a funny weatherman to get the news across . . . then I can't say they're doing the wrong thing."

Ernie Schultz is also quick to bid good riddance to the journalism of the 1960s. He, too, doesn't miss "the Olympian presence in the studio with the deep voice reporting on what government did." Local stations, Schultz says, respond to the people by providing coverage of government in terms of its effect on them. TV news is "the last mass medium," and health as well as consumer information is "the information we found that the people regarded as news."

HAPPY TALK

Any discussion of popularization of news has to include the concept of happy talk. That's the anchors' seemingly ad-libbed, folksy-friendly chatter among themselves. Happy talkers thank reporters on the air for doing their job. Happy talkers wedge in all sorts of personal minutiae, ranging from last night's Lamaze class to this morning's golf score. And happy talkers bring everybody—on-air personalities and viewers—together into one big happy television family, sharing the good and bad, in news and commercials. With that goes a lot of on-camera giggling and meaningless chitchat but, most important, viewers get the feeling that they are one of the boys or one of the girls.

News executives like Schultz maintain that people respond favorably to the interrelationships on the anchor set. "It (happy talk) increased viewership . . . trust, respect, and credibility." As for newspaper critics who zero in on happy talk, he says they "couldn't report their way out of a paper bag."

According to Schultz, station owners perceive journalists as purists who have trouble facing the reality of economics or even reading a budget. In other words, journalists ought to understand "we're in the business to make money."

THE PLUS SIDE OF THE JOURNALISM LEDGER

Some journalists look at show biz and journalism as not necessarily mutually exclusive areas, and they are dedicated to accomplishing both. They seek out and use techniques for making news delivery on any subject both interesting and exciting. They don't assume economic or political stories have to be run with deadly **talking heads** only. They know that there is a human ingredient in every news story. Each key player—the news director, producer, writer, reporter, camera operator, and editor—work together toward a common goal. As members of a team, they use all the available tools of picture and sound to tell the human story—complete with beginning, middle, end, *and* credibility—in an entertaining way.

These movers and shakers have continued to condemn any compromise of journalistic standards because of ratings pressures. They do it by voice and by deed.

Each year, *Broadcasting* compiles a list of the serious investigative reports and documentaries produced by news departments that range in market size from the smallest to the biggest. The following is a typical sampling.

- Low birth weight deliveries: KPNX-TV, Phoenix
- Illegal dumping of toxic wastes: WJW-TV, Cleveland
- Unethical car dealers: WXFL-TV, Tampa-St. Petersburg
- School bus safety: KWTV-TV, Oklahoma City
- Police scandals: WKYC-TV, Cleveland
- Transit police scandals: WNBC-TV, New York

- Teenage violence: WUSA-TV, Washington
- Black aviators: KRIV-TV, Houston
- Violence against Asians: WNEV-TV, Boston
- Laboratory testing mistakes: WRC-TV, Washington
- Political corruption: WTVH-TV, Syracuse
- Unsafe cars: KMGH-TV, Detroit
- Nurse shortages: WWSB, Sarasota
- Skyscraper safety: WJXT-TV, Jacksonville

All Post-Newsweek stations, Jim Snyder says, aim for courageous, tough, in-depth reports, not sensationalism. He points with pride to another WJXT-TV news special, "Smell the Mud," which he credits with causing the enactment of Florida's toughest local ordinances on air pollution.

To Bob Mulholland, emphasis on ratings means emphasis on making money, and too many stations make judgments "strictly on how much money they're making." However, Mulholland, who worked for years on the commercially and journalistically successful Huntley-Brinkley show, also feels a news product can be professional *and* make money.

For many news operations, the good old days are still here. In the golden age of broadcast journalism, the 1950s and 1960s, it was mandatory for writers and reporters to have a solid newspaper background. NBC News, for example, required five years' experience working for metropolitan newspapers. Management's reasoning was that these writers and reporters knew how to dig for a story and had been drilled with high ethical and professional standards. *Writing* was the heart of newscasting to the extent that staffers would never go on to reporting, producing, and executive positions until they had mastered that art. It also meant that newscasts were literate, understandable, cohesive, complete, accurate, and impartial, as well as colorful.

THE STANDARDS AT KING

The King Broadcasting Company's dedication to broadcast journalism follows similar standards today and, as a result, its product is legendary. For one thing, KING-TV in Seattle believes in serving its audience 52 weeks a year and does nothing differently during ratings sweeps. At the core of viewer loyalty, according to Donald Varyu, "are the things that are perceptively the simplest—writing, good reporting—the things that frequently get overlooked for the new technology." Varyu says he spends most of his time perfecting the craft of writing, which "may be archaic, but I still believe in it." At KING-TV, the emphasis is also on depth over item count.

Varyu admits to possibly erring on the side of playing God and deciding what people need to know, a basic mission of telling them "what they don't know about rather than what they do want to know about."

At KING-TV's sister station in Portland, the dedication to good journalism is equally strong. Reagan Ramsey regards KGW-TV as "a positive element of change in the community." If Lincoln Steffens, the famous journalist who exposed political corruption in the early 1900s, were alive today and muckraking with ENG instead of a pencil, he would be working for KGW. Staffers follow stiff rules about no titillation coverage and no routine murders, house fires, or car wrecks. They don't consider preying on individual misery as news and occasionally miss a story because of their policy of not covering the sensational. KGW producers do mini-docs when the time is right, not because the ratings race is on.

The philosophy comes from the top, King corporate management. As Ramsey puts it, "At the end of the year, when they judge my performance, they say, what did you do that's different? They never say, did news make money, what are you going to do to increase the profitability?"

Not that there isn't ratings pressure. Management sees nothing inconsistent in having an impact in the community and also getting people to watch. At KGW, the attitude is that journalism is a profession with the same responsibility as a physician or attorney. As with a good doctor or lawyer, the dedication is to improving life rather than grinding out as many cases as possible.

Ramsey's stance is a firm one, even when the news department's commitment creates havoc with King company profits. In one instance, some Portland radio stations, including the King company's, were broadcasting a series of commercials touting a car that didn't exist. The reason, according to Ramsey, was to revitalize the then-depressed radio commercial business. "We heard about it, documented it, proved it, and followed it up with the attorney general. Our two radio stations heard about (our investigation) and they screamed bloody murder." It wasn't long before corporate management pointed out that two of the company's revenue-producing properties were getting hurt by the investigation. "I said it's journalistically irresponsible for us to ignore this story. If it damages you guys, you shouldn't have been doing something dishonest." The two radio managers visited the newsroom. "They said, 'You're going to kill us. You can't run it.' We said tune in at five and we ran it."

PUBLIC SERVICE AT WTVT

In Tampa, WTVT's philosophy is also built on a core of public service. As Jim West puts it, "I know that may sound trite and we talk about this business being built on the public trust, but it is." At WTVT, being number one doesn't necessarily mean always delivering to the viewers what they want. "Sometimes you . . . give what they need to know because the viewer may not always be smart enough to know what they need to know."

WTVT, like KING and KGW, no longer bombards the viewer with mini-docs during sweeps. For the CBS affiliate, mini-docs have become more a year-round commitment. Included in that year-round commitment is a daily "Closer Look" segment, a four- to five-minute story taking the most interesting, most important story of the day and giving it added treatment. WTVT's emphasis is on delivering important knowledge interestingly by addressing the viewers most affected and utilizing production values such as graphics, charts, and maps.

THE MICROCOSM APPROACH

The people-oriented approach sounds obvious, but few producers know how to use it. One effective technique for humanizing coverage is to epitomize a big story by taking one aspect of it and telling it completely. The subject of crime is a good example, particularly on Mondays, after a typically bloody weekend in most large cities. The tendency is to run the footage of bundled bodies, bloodstained sidewalks, and bullet-riddled walls, one story after the other. The result of this approach is that the viewer learns little of substance about any one homicide while becoming increasingly immune to violence. An alternative is to pick just one murder case from the police blotter and tell it well, to humanize the victim for the public. This microcosm approach, which zeroes in on the victim's family, friends, neighbors, profession, habits, and hobbies, serves another vital purpose. It turns what may be cynically referred to as a cheap murder into real tragedy.

However, not all news executives have free reign, and fighting a materialistic environment can be frustrating. KSTP's Larry Price is among those who miss the good old days. He points to the present-day insensitivity to viewers and the emphasis on marketing as what's wrong with TV news. Price's ideal world is the one of the 1960s, "where you could just go out and do a very good newscast and let it survive on its merits, but it doesn't work that way" now.

KRON-TV's Herbert Dudnick is from those 1960s, and he tries to maintain the same standards. His product trails in the ratings race, but Dudnick sticks to his philosophy of being more issue-oriented than number one KGO-TV. Dudnick also does nothing special during the ratings sweeps, because he already has a daily special segment.

In the 28 years Dudnick has been in the news business, he too has seen the change away from newspeople determining what the viewer shall receive. He now feels pressure, for example, to make sure a story with some sort of emotional hook hits toward the beginning of his 6 o'clock newscast.

A MIDDLE GROUND

For several news directors, there is a delicate balance between what the audience should have and what it wants. KSL-TV is among those that recognize the reality of ratings but insist on competing in good taste. J. Spencer Kinard admits he'd be happy never to see another rating book. Yet, he believes in competition and thinks it has been good for the industry and the community. Kinard stresses accuracy, fairness, and speed. Having people watch his news is also important, so he urges his news people to be interesting as well.

In addition, the Salt Lake City veteran pins most of the responsibility for a good product on the producers rather than the viewers. Kinard acknowledges the public's obvious influence on the ratings books, but thinks "we, too, have had an opportunity to create an appetite among the pubic. The public didn't know they wanted a report on education until we did it." As for sleaze and tease, he doesn't buy the idea that viewers force them onto television. "If we didn't give it to them in the first place, the public wouldn't ask for it because they wouldn't know it existed."

James L. Boyer wants to be number one *and* best. He wants KOMO to be content-oriented, relevant, understandable, and with just enough showmanship for Seattle news consumers.

As for the 1960s, Boyer says he's never heard anybody talk about the bad old days. "Who the hell is Fred Friendly (former CBS News president) to say automatically this is what 200 million people ought to know? The opposite of that, he is saying, is what he doesn't report he doesn't want them to know. They don't deserve to know. I think that's arrogant." However, Boyer denies falling into the let-the-people-have-anything-they-want-camp. While news is obligated to report what the public ought to know, "we also have to know what the public is interested in knowing more about."

Do it with style is what Boyer is saying, because the key is still getting as many people as possible to watch. As a result, KOMO news does lay it on for the sweeps, airing series aimed at generating an audience. However, Boyer insists, they are not "flash and trash." KOMO has done sweeps mini-docs on street gangs and teenage runaways as well as some less serious, but nontitillating "fun stuff."

Reid Johnson of WCCO in Minneapolis also considers it arrogant "to determine your agenda for the community." Still, he acknowledges, "there are some things as journalists we have to do to get people thinking about things." WCCO is also heavily community-minded, which is a clearly growing national trend. The Minneapolis station broadcasts from the state fair, major high school sports tournaments, the St. Paul Winter Carnival, and a variety of other civic festivities. As for sweeps programming, Johnson insists his station stays out of the gutter. His mini-docs are on self-help subjects such as how to finance college.

Jeff Rosser also reaches for middle ground. His user-friendly approach to the news is a little more casual and less formal—a compromise between serious traditional news and happy talk. "People communicating with people," is the way he puts it. Rosser has no use for the pedestal status of news in the good old days, but he doesn't like happy talk. "Our job is to cover the news, to find out what happened today and what it means to all of us . . . to our neighbors, to our friends and our family."

THE FOREIGN MARKET

The ratings frenzy is an American export that is becoming increasingly visible in Japan, Italy and France, but so far it hasn't reached Sweden or Great Britain.

Swedish TV, for example, is under no pressure to increase ratings since both of the country's channels are part of the Swedish Broadcasting Corporation. Janne Andersson, chief of Channel 2, the regionalized outlet, admits he would love to see the figures anyway, but he doesn't have the budget to pay for a survey.

Other Swedish TV executives are starting to think more like U.S. news entrepreneurs. Network national news chief Ingemar Odlander, for example, wants people to watch when he does something important. Unlike the regions, Odlander gets ratings of his half-hour nightly

"Rapport" newscast. Odlander reports a comfortable score of at least 2.5 million viewers out of a population of 8 million.

The BBC, reports Rick Thompson, is not as ratings crazy as the United States. There is pressure, but it's not as intense as in America. The difference, Thompson says, is that their budget does not depend on the number of viewers. Like many of his counterparts in America, Thompson sees no real fundamental problem with making news popular and making it responsible.

Not that the Beeb was always that professional. According to Chris Cramer, BBC journalism has become more responsible in the past four years, having gone through its sensational period. For years, Cramer says, the two competitors, BBC and ITN, had been covering the same stories in the same populist way. Actually, Prime Minister Margaret Thatcher is credited as the primary reason why coverage was cleaned up. She has been determined to crack down on all violence and sex on TV.

While the BBC insists ratings are not a matter of life or death, its competition in Scotland, Scottish TV, is proud of its position in the numbers war. Boss David Scott boasts a maximum viewership of close to 2 million, four times as many as BBC in Scotland. However, Scott feels broadcast news there will never become too popular or too sensational.

Liz Forgan, like so many foreign TV news executives, is schizophrenic about ratings. Is she concerned? "Of course, I am a bit, and the ratings of the Channel Four news are a bit disappointing." Her erudite program is up against the most popular shows in England, and the best she can hope for is 2 million viewers against the other news programs, which get between 9 million and 12 million. In other words, despite the intent to leave Channel Four's alternative newscast immune from commercial and ratings pressure, its executives still worry.

As TV journalists abroad start worrying about ratings, they also begin sharing another import from us, the resulting impact on costs. As competition intensifies, so does the need to spend more money. The popular on-air talent gets expensive, raids on other companies become common, and the budget to cover the more promotable news stories explodes.

According to Elie Vannier, the only serious French competition exists between the recently privatized channel and his state-owned Antenne 2. Vannier says the privatizing of the competing channel cost him a fortune. Bidding for sports events jumped the price 10 times, and movies, which used to cost no more than 2.5 million francs, now range up to 10 million francs.

The drain on news budgeting is equally serious, Antenne 2 now paying considerably more for anchors, reporters, and crew members. However, Vannier believes in competition and, as president, was toe to toe with the privatized channel. Nevertheless, he has never thought news should be popularized for the sake of ratings. "Emphasis being put on anchormen is wrong." Yet Vannier does not think the news executive should be holier-than-thou. "I think you have to give people what they want or they won't watch you. I hate a reporter or journalist who says, 'I know better than they do.'"

Italy's RAI is also showing increasing concern about ratings. Despite the emphasis on government control, shows are judged by the Auditel ratings system, a meter device. A third of Auditel, or AGF, is owned and controlled by RAI, the rest split between private groups and advertising agencies.

Jian Carlo Pizzirami, the news planning and coordinating chief for RAI's Channel 3 in Rome, admits he is concerned about ratings, but not at the expense of good programming. "We are interested in getting as many people as we can, but we cannot sacrifice. We have to give the culture side."

While RAI 2's chief news editor, Vittorio Panchetti, claims ratings are not a problem for news executives, he is very aware of the standings. Panchetti says RAI 1 is first, RAI 2 second, and RAI 3 third. On a typical September day, the first channel, the one identified with Christian Democrats, had 7 million viewers; RAI 2, which is identified with the Socialists, had 5 million; while the Communist-influenced regionalized channel lagged behind, with a million viewers.

In Japan, the NHK's Yoshihiko Kubota cites a policy that does not allow sensationalism or propaganda into newscasts. However, the newscaster admits, "We do have to think about how to make a story more interesting." Kubota says this is not the same approach as that of competing commercial stations, which try to make crime stories more entertaining. NHK's news practices have changed dramatically in 10 years, swinging away from accidents and criminal cases to follow-up and investigative reporting.

But ratings are an issue for NHK. Producer Kenichiro Iwamoto says the network is "concerned to some extent, but probably not as much as the commercial stations," despite the absence of sponsors and commercials.

If there were a handy Japanese phrase for *déja vu*, Iwamoto would have used it. He says commercial stations in Japan, like American affiliates in the 1970s, are realizing there are profits in news. Until recently commercial networks didn't emphasize news because it wasn't getting very high ratings. But, Iwamoto says, Japanese are becoming addicted to news, and commercial stations are reacting accordingly.

Kazuo Hori, in NHK's news operation in Shizuoka, confirms the network's worry about the number of viewers. An interesting program means nothing, Hori says, unless people are watching. For Hori this is a problem. He points out that NHK was established 30 years ago for the purpose of providing good quality. "But now we also have to think about the number of viewers. We are in a dilemma to take more viewers or quality." The impact on viewers is the most important factor, says the news vice president. "We have to think always our news must affect the people in a good way, not bad ways. We have to always think that we are on the viewer's side."

Some Japanese news executives are reluctant to admit their operations are being affected by the pursuit of viewers. In Fukishima, station TUF insists it doesn't let ratings affect its news coverage. Ratings are also not a factor, they say, in hiring newscasters. However, an official in TUF's parent network, the Tokyo Broadcasting System in Tokyo, recognizes the change in emphasis and is concerned about it. In Japan, as in the United States, newscasts at one point were not concerned with generating money. But, according to TBS's Junji Kitadai, "Now it is generating money, so there has been a change of awareness, in a way, on the part of the broadcasters and the sponsors and also the audience."

As in America, this new awareness has created a more competitive news product. Kitadai, recognizing the danger, sounds like many U.S. journalist soothsayers in the late 1960s: "People like me who believe in the old tradition of journalism regret [it] very much, but it is true that more and more entertainment or show biz appearances [are] coming in the news."

His words could have come from any number of news directors here. "We still believe in news tradition. But in order to have good ratings, which are demanded by the management . . . there are more and more cosmetic things which apply to the presentation of news, not the content." Kitadai talks about the increasing use of "announcers or even actors and actresses" hired as anchorpersons. He talks about NIPs and Media Research, the Japanese versions of the Nielsen and Arbitron ratings services. He talks also about the increase in promoting and hyping news. "Maybe we should do that more. I just don't know. It's supply and demand. Japanese broadcast journalism is getting more and more like the United States."

PROJECTION

The hard reality for U.S. television news is that it will have to continue making money. This is particularly true in a deregulation environment in which TV stations and networks are bought with little cash and when the primary motive is profit. There already exists ample evidence of new ownership whittling away at, or in some cases completely eliminating, news operations.

As long as news is an integral part of the moneymaking machine, it will suffer consequences accordingly. In Los Angeles, for example, KCAL-TV broadcast journalists face the humiliation of their own parent company promoting its entertainment programs on other channels, in competition with its own newscast. KCAL Channel 9's new owner is the Walt Disney Company. A Disney series, "Hard Time on Planet Earth," airs on KCBS-TV Channel 2 at 8 P.M., the same time as KCAL news. To the chagrin of Channel 9's news staff, Disney sees nothing wrong with promoting the Channel 2 series.

For Disney, the buck starts anywhere—as it does for many other profit-oriented news operations born in the news-heavy turmoil of the 1960s. When the 1970s station owners and managers learned how to take advantage of the nation's thirst for news, winning the ratings race became everything. Getting them into the tent became the priority, and the barkers' pitches, the promos, outsleazed the titillating news product itself.

One trend to keep an eye on is the tendency to place non-news-people in charge of news programs. ABC set the precedent by moving sports executive Roone Arledge into that position. In 1989, NBC jumped on the bandwagon by naming Dick Ebersol, its sport president, to oversee "Today," which had been traditionally controlled by the network's news department.

On the positive side, there is a surge in emphasis on meaningful mini-docs and investigative reporting at the local level, with more stations producing the pieces year-round rather than only during ratings sweeps. The programs are entertainingly produced and enthusiastically received by the public, proof that show biz and journalism can mix.

It is also possible the sensational popularity of the Winfrey and Donahue type of live, audience-participation show will have a positive effect on future news programming. KING-TV's Varyu is among news executives who see a future news program vehicle in them. Varyu calls it the town meeting type of formula, an expansion of audience, anchor, and newsmaker. Audience participation programs based on contemporary significant themes are already being aired on British and Swedish TV and radio.

The growing core of dedicated broadcast journalists, coupled with an awakening public, could do it. Together, they could generate enough power and professionalism to propel the industry forward to a positive, innovative decade. It could be an era that future general managers won't dare describe as having "nothing to do with journalism."

6

---□---

THE NEWS MEDDLERS

Commercialization
of TV News
to the Extreme

The pursuit of popularization takes an inevitable turn from the "give them anything they want" syndrome to bent rules, tainted standards, and the ultimate cheat, ersatz news.

Electronic journalism isn't the only medium to be catalyzed by high tech and emerge radically and forever altered during the latter decades of the 20th century. Its distant cousin, advertising, has also been energized into a tremendous force in society. However, they haven't exactly been kissing cousins, journalism and advertising. At best, the relationship has been a schizophrenic one as far as news is concerned. Both electronic and print journalists have always looked on advertising as a sort of Daddy Warbucks: always there with the cash but far enough away that the recipient doesn't get tainted by the source. In other words, the Little Orphan Annies from the editorial side traditionally have taken the money and run from any association or suggestion of influence from the advertising side.

NARROWCASTING

But along came sophisticated computers and a socioeconomic marvel called market research. It has become a major tactic advertisers use to influence the news product their way. And so far, nobody has come up with the right weapon to keep this particular villain at bay.

Instead of which toothpaste consumers prefer, this process determines which news segments viewers want. The audience's desires then become the criteria, the determinants of what goes into a newscast. The inevitable result is called **narrowcasting**—designing the news product for specialized, segmented portions of the population. A program that hits the air early in the afternoon, say 4 P.M., will be heavy in stories of interest to older people, shut-ins, housewives, or blue collar workers who get home early. Later newscasts will lean toward younger, more affluent people, particularly women, considered to be the household's primary buying force.

Gary Cummings, former general manager of WBBM-TV in Chicago, describes the process for the Gannett Center *Journal*:

> Slowly at first, then with greater speed, the research locomotive took advertisers to the land of demographics—specific age, sex, economic groupings. Research told advertisers it wasn't *how many* eyeballs they reached that mattered, but *whose* eyeballs they were.

Advertisers want to reach the specific person who will buy their perfume and that eliminates 55-year-old blue collar men. If the product is shave cream, forget pubescent girls. As time goes by, the demographic pie is cut into thinner and thinner slices. It's a matter of "identifying buying segments such as 24-to-34-year-old women, and encouraging advertisers to reach their target audiences without wasting money on audiences not interested in buying their particular product," says Cummings.

On the network level, watch the news stories and note the commercial content of the morning shows from about 8:30 A.M. to 9 A.M. Take your choice among the latest fashions from Milan, a demo right in the studio kitchen on fixing Thanksgiving dinner, or the latest on breast cancer.

Other ethical problems, in addition to blocking out portions of the news budget for the segmented audience, also arise. Specialized news sections or programs, for example, are sponsored by companies producing related products. Brokerage firms pay for the business report, pharmaceuticals for the health section, department stores for fashions. Narrowcasting is designed to create happiness for everybody—advertisers, station managers, and viewers. Everybody except for that lone soul concerned about credibility.

As for racial, sexual, and social harmony, this trend represents a step backward. News programs aimed exclusively at a specific race, religion, sex, or social group can fuel polarization.

VIDEO NEWS RELEASES

Why not take commercialization of news to its natural absurdity—that is, skip the middle person and let the advertiser produce news packages? The public relations companies have already answered the question. The **VNR, video news release,** is one of the hottest items in public relations today. Companies are producing and distributing, by satellite, taped, slickly packaged electronic releases that look like news reports. The VNRs, with the necessary visuals, graphics, interviews, and narration already in place are timed to slide into any newscast.

As for the ethical aspects of the mushrooming practice, consumer guru Ralph Nader has already taken up the challenge. Nader wants the Food and Drug Administration, the Federal Trade Commission, and the Federal Communications Commission to investigate VNRs.

In the meantime, the electronic press release flourishes. Thousands are distributed each year and one company, Medialink, brazenly puts out an annual list of its "Top-10 VNRs." Medialink's number one VNR in 1988 was for the legally embattled stock brokerage firm of Drexel Burnham Lambert. It was seen by more than *80 million* viewers on network, cable, and local television newscasts across the country. The Drexel VNR was just one of 800 distributed in one year by one company. In 1989, Medialink went international, producing and distributing a VNR for Virgilio Barco, the president of Colombia. The release, in which President Barco asked Americans to stop using cocaine, was seen 75 million times.

ERSATZ NEWS

When it occurred to production companies that they could go all the way and thoroughly entertain the viewer, they came up with ersatz news. The concept is referred to as **trash TV** or **tabloid TV** and it unashamedly caters to the audience. Its story could almost be told in headlines. "Here Come the News Punks," proclaims *Channels* magazine. "It's a Crime What They Offer to TV" is the headline on media critic Howard Rosenberg's column in the *Los Angeles Times*. Other newspaper headlines spark optimism in the fight against trash: "KABC to Drop Morton Downey Talk Show," "'USA Today on TV' at Ratings Crossroads," "CBS' 'Inside Edition' Puts a Chill on TV's Hard News Coverage," "The Heat Is On: ABC Drops Two Tabloid Specials," and "Some Advertisers Shying Away From Trash TV Programs."

The question is how many advertisers will shy away? More important, will trash TV fade as fast as it emerged? The answers are related because the product, distributed to individual stations through a nonnetwork method referred to as syndication, will last only as long as it is profitable. In the meantime, the programs proliferate like the bad seeds from outer space in science fiction movies. They include "GERALDO," "The Reporters," "A Current Affair," "Inside Edition," "Unsolved Mysteries," "America's Most Wanted," "Hard Copy," "Crime Diaries," "Crimestoppers 800," "The Investigators," "Crimewatch," and a slew of others ranging from "Has Anybody Seen My Child?" to "Cop Talk: Behind the Shield."

Some of these trash TV shows spawned by Morton Downey Jr. and Geraldo Rivera are on the air, some get canceled out by audience and advertiser revulsion and others are in the pilot or planning stage, awaiting station acceptance.

In any event, the means are the same (sensationalism) as the end (money). Sadly, hitherto solid shows like "Oprah Winfrey" and "Phil Donahue" have changed directions accordingly. In the face of the competition and trend, the themes of their programs have become increasingly sanguinary and sensuous.

THE ZIPPING OF THE DOWNEY SHOW

When enough of Downey's 28 million viewers finally got fed up with tactics bordering on incitement, they stopped watching. His ratings dipped, advertisers got nervous, and the program was finally canceled in July 1989. In other words, "The Morton Downey Jr. Show" died when it stopped being profitable. This despite Downey's attempts to assure affiliates by letter that he would be less abrasive to guests and would eliminate "harsh language across the board." But Downey himself hasn't been zipped. From the other side of his mouth Downey spouts other goodies, in the form of a record album called "Morton

Downey Jr. Sings." Some 500,000 copies of the potential gold record have gone to stores, giving exposure to his venom in another medium. With musical accompaniment, he trashes dope dealers as well as doctors, lawyers, politicians, and foreign investors. The drug dealer he refers to as "Ya slime suckin', drug pushin' S.O.B." who he hopes will "die slow." His refrain for doctors, "Operate, operate, gotta pay for my Mercedes-Benz."

Among the many journalists who see danger in trash TV is NBC's Garrick Utley. He told the 1988 convention of news directors to protest more actively the antics of **junk news**. The reporter is concerned that "their very presence on the television screen gives the subjects a certain legitimacy, that theirs is proper behavior." Utley called the programs "entertainment posing as news, as documentaries, as journalism." Unfortunately, advertisers and programmers are also aware of the identification and purposely schedule the junk shows close to legitimate news programs.

To the *Washington Post*'s media critic, Tom Shales, the emphasis on crime also presents serious societal problems. Shales, in his February 5, 1989, column, charges that it exaggerates fears people already have about crime in the outside world. "It advances the idea that brutality and killing are normal and routine components of everyday life, and it decreases the public appetite for serious news that really matters."

Trash TV is an inevitable extension of a philosophy built on greed at the expense of the journalism profession, viewers, and society as a whole.

STAGING THE NEWS

One of the dominant characteristics of the junk news shows is their technique of reenacting real events with actors. Although real journalists don't, or rather aren't supposed to, reenact or **stage** news, the problem has never been completely eliminated from broadcasting.

Even the exalted BBC is not exempt from the practice. "We frequently reconstruct shots," admits the Beeb's Chris Cramer. If his crew is doing a sequence involving a man receiving a letter from a postman, for instance, it is not unusual to have it redelivered in order to cover the action from both angles. There are also occasions in Great Britain when sound effects, like burning fires or sirens, are dubbed onto TV news footage, another technique frowned on by most U.S. news executives. Cramer, however, says BBC does not reconstruct demonstrations or other news events.

In Japan, TV Asahi was caught staging a gang fight in 1985. That fakery led to the conviction of several broadcast executives, the suicide of a participant's mother, and heavy negative publicity. Although a hard news program was not involved in the scandal, TV Asahi news now follows a particularly strict policy and makes an all-out effort to get more in-depth news coverage.

Reenactment in what is considered to be legitimate broadcast journalism in the United States occurs to varying extents. An American TV crew, for example, might also be tempted to redeliver a letter. Other borderline practices also stubbornly continue, such as networks and stations continuing to use the **reverse question** tactic in interviews. This practice entails reshooting the reporter's questions after an interview is completed and after the subject has left. The object is to make the reporter look good, smoothly asking the questions in full frame for the final edited story on the news. Another related and questionable technique is the phony reaction shot. That involves showing the reporter reacting to an answer although the subject is no longer there. Since the movie *Broadcast News* highlighted William Hurt's character staging tears in a reconstructed reaction, however, the public has become sensitive to that ploy.

The more serious staging problems in the United States tend to fade in and out. In the turbulent 1960s some TV crews were known to have political demonstration participants spring into action at rallies staged strictly for the camera, but that practice has declined.

However, using actors to reenact news and historical events has become commonplace in the syndicated tabloid shows. It is a precedent set on the entertainment side of television and cinema with the so-called "docudrama" approach. The story of a major event, such as the Watergate scandal, the Oliver North case, or the murder of the civil rights workers in Mississippi in the 1960s, is acted out in movie form. In most cases, the dialogue is fabricated and, in some cases, characters are created for the convenience of the story.

In 1989, two new network news shows caught in the race for ratings crossed the line and introduced heavy use of the reenactment technique. By the end of the year, fortunately, NBC's "Yesterday, Today & Tomorrow" and CBS's "Saturday Night with Connie Chung" reacted to criticism and stopped using re-creations. However, both networks managed to keep the stage door opened. NBC transferred "Yesterday, Today & Tomorrow" from its news division to its entertainment division, which portends a possible revival of reenactments. As for CBS, by the end of the year it had not issued a clear-cut policy statement on the subject of acting out the news.

One instance of reenactment on a network newscast backfired. This was on July 21, 1989, on ABC's "World News Tonight." The news department simulated photographs of an alleged spy handing over a briefcase to the Russians. However, the program failed to label the realistic-looking photos as a reenactment, and anchor Peter Jennings had to apologize for the error on a later newscast. In the July 26, 1989, *Los Angeles Times*, Richard Wald, ABC's senior vice president of network news, defended staged reenactments as nothing more than extensions of "courtroom sketches." Two other networks, the Columbia Broadcasting System and the Public Broadcasting System, admitted they use reenactments, but are careful to label them as such. Cable News Network's executive vice president, Ed Turner, told the *Times* that CNN never dramatizes any news event. Turner said, "That's what a reporter is for and what good writing is all about. There is already so much blending of the 'docu' and 'drama' and the blurring of the lines between news and entertainment, that I fear increased use of this practice is dangerous."

THE CHEAP SHOT

The bring-'em-into-the-tent—or tease 'em into viewing—syndrome takes its toll in still another dimension, this one involving people thrust into the news innocently and tragically. It involves the too frequent practice of broadcast journalists badgering dazed relatives of victims struck down by plane crashes, murders, or other catastrophes. In the classic case, the distraught man sits on the curb in shock, his burned-out house the backdrop. The young, aggressive TV reporter, looking for on-camera emotion, preferably tears, asks him, "How does it feel, losing your wife, three kids, house, and all your belongings in that fire?" It happens.

A real victim from New York, whose son died in a plane crash, wrote to *Time* Magazine on March 27, 1989:

> I found the behavior of the national press at the airport terminal on the night of the tragedy reprehensible. Within two minutes of discovering that my son was dead, I was swamped by reporters as a security officer attempted to escort me to privacy. My passage was further hindered by a journalist who planted his foot on mine to prevent me from moving. Then came the ultimate question: "How do you feel?" Have new words been invented to describe the anguish every family member experienced that night? Journalists became part of the problem.

Other, more compassionate U.S. news departments cover tragedies responsibly. In the same issue of *Time,* a Minnesota woman wrote:

> You noted that station WCCO-TV in Minneapolis forbids its re-
> porters to ask victims' relatives how they feel. I applaud this
> emerging trend. In 1986, my brother drowned. WCCO covered his
> death and funeral in a responsible manner and with concern for my
> family's wishes.

Overseas, Thames TV makes a major effort to avoid invading the
privacy of people involved in a news story despite the British police
practice of providing to the media the names of relatives of crime
victims. Japanese broadcast journalists also take their ethics seriously.
At NHK, for example, staffers do not interview parents of a kidnapped
child. They will also honor the customary police request not to report
a kidnapping while it is still going on.

PROJECTION

As blatant commercialization metastasized in the 1980s, it infected
more and more of society with its far-reaching tentacles of narrowcasting, video news releases, trash TV, and the revival of staging.

Whether broadcast journalism ethics will flourish in the decade of
the 1990s depends on how many Reagan Ramseys and Donald Varyus
stay in the business. It also depends on the even more powerful force,
the audience. And it depends on the people who represent their
interests, such as responsible media critics like Jeff Greenfield, Howard Rosenberg, and Tom Shales. With their strong voices raised in
protest, along with those of responsible vocal TV journalists and caring
educators, the public has a fighting chance of becoming even more
aware of the journalistic damage being done. Specifically, the journalistic damage is reflected in a tendency on the part of viewers to confuse
news and entertainment.

Signs of viewer enlightenment are already showing up in letters,
polls, ratings, and the TV junk show graveyard. A Roper Organization
survey reports a drop in the number of people who consider TV their
sole source of news and who feel television is the most believable news
medium. The latter is down by 6 percent over a 2-year period, while
the number of people who consider newspapers most believable is up
by the same amount. However, a 4-year survey conducted by the
Gallup Organization for the Times Mirror Center for People & the
Press reports a decline in public confidence in both electronic and print

media. The study, concluding at the end of 1989, shows that all three network news operations—CBS, NBC and ABC—are suffering drops in believability. CBS has the biggest problem, with an 8 percent drop. CNN, which concentrates more on straight reporting and less on interpreting, comes out second best in believability among all news organizations. The *Wall Street Journal* is first. Generally, though, the Gallup poll contradicts the Roper survey by finding that daily newspapers, not TV, are losing the most ground in public confidence.

As for narrowcasting and video news releases, it is unlikely those forces will taper off. Market research isn't going away and news releases will continue proliferating as they did well before television news. Hopefully, however, these and all the spinoffs from the materialistic side of journalism will be policed with the same kind of enthusiasm that paces the ratings race itself. But, if broadcast journalism insists on enlisting the help of Hollywood's central casting to help deliver the message, it will need a great deal of help in separating fact from fiction.

7

NEWS PROGRAMMING

The demand for local news is bigger than ever, but dramatic changes are taking place as to when newscasts go on the air and their length. It's a matter of keeping up with the new look in viewers' lifestyles, competition, technology, and the broadening news menu.

News junkies, those dedicated fans who never seem to get enough, continue to rule the airwaves at home and abroad. Nevertheless, the popularity of newscasts in the traditional time periods may be on the wane. Indeed, although the polls indicate gains for the familiar early evening local TV news, it wasn't long ago that other polls had Tom Dewey beating Harry Truman for the presidency. Although the early evening block won't wake up one morning to find itself completely defeated, as did Dewey in 1948, early-rising viewers are awakening to a tremendous increase in the amount of news given in the dawn hours. As for that old-time, late night favorite, the 11 o'clock news, the portent is ominous.

At best, programming is mercurial, depending on viewer habit, technology, the economy, and profitability. As a result, programs face a continuing challenge to adapt. For the time being, at least, these forces are creating a trend toward more local TV news, but in different time periods.

THE LATE AFTERNOON BLOCK

A Television Information Office (TIO) survey is an example of the fact that polls can be misleading. It shows a slight trend toward expansion of newscasts in the traditional late afternoon news period. However, the basis of its conclusions, the raw numbers, are not impressive and show that the *rate of expansion* has actually slowed. Most news executives interviewed agree that viewers are serious about watching less during the twilight zone and more during other time periods. At the same time, there is ample evidence that producers are interspersing an increasing amount of soft news, features, into the late afternoon period.

As for the numbers, TIO reports that more than twice as many network-affiliated stations increased their early evening local television news programming as decreased it. The TIO survey, based on A. C. Nielsen data covering 627 affiliates, was released in July 1988. Its findings:

- Two hundred and sixty-five affiliates broadcast one hour or more of local news between 4 P.M. and 7:30 P.M., *a mere 2 percent increase over the previous year.*
- Forty-nine affiliates increased local early evening news programming, but *24 others reduced it.*

According to the TIO-Nielsen survey raw numbers, only five additional stations carried early evening local news and only two of those were network affiliated. Only six stations increased to an hour and three to 90 minutes. Two affiliates actually cut back from two hours, one from two and a half hours, and there was no change in the three-hour block. Only one polled station programmed the latter. Forty-seven affiliates added a half hour, but only two added one hour, and 24 dropped a half hour.

News directors interviewed for *Changing Channels* see major changes under way but report that the bulk of their product still currently airs in the early evening period, usually sandwiching network news. This is in line with the national trend, which sees some 140 stations now airing 30 minutes of news before and after the networks'. More significant is the tendency to upstage the networks. Almost all stations with one-hour newscasts and most with 90-minute programs air them *before* their network's showcases.

The troublesome aspect of this trend involves the local stations' newly acquired technological ability to cover national and international news. In some cases, the affiliate can help the network by only touching on the non-local stories and teasing the more expanded version coming up on the network newscasts fronted by Tom Brokaw, Peter Jennings, or Dan Rather. However, the temptation is too much for local stations. Many of them end up outtelling and outdoing the networks by providing their viewers with the complete story regardless of its scope. The time difference also works against the networks in that, in many cases, the big shows are fed into the stations from New York, taped, and played on the air hours later. Local news departments feel any story breaking after the network show has been taped is fair game. It doesn't help that the network hierarchy is still too slow to update breaking stories for the Midwest and West time zones. Junkies or not, viewers are unlikely to hang in for an outdated newscast if they have already heard and seen it all.

Alternative Newscasting

While many stations design the 5 P.M. newscast with all the news of the day for a "newscast of record," Jim West takes a different tack for WTVT. He doesn't throw all the national and international news available to him into his 5 P.M. show. Rather than sabotage the network news West takes an offbeat approach. The theme, customized for Tampa viewers, features a lighter approach, with emphasis on human interest, entertainment, and recreational leisure fitness. "It's what the average Joe Citizen is doing with his free time," explains West. Instead of major league baseball, football, basketball, or hockey, viewers get a sampling of sports in which they actually participate, like sailing, fishing, boating, wind surfing, and archery.

As for hard news on the 5 o'clock, there are only two five-minute packages in a structure dominated by consumer lifestyles, fashions, dining, the arts, and family health. West's 6 P.M. newscast runs with a harder edge, emphasizing news peppered with investigative reporting and special projects.

West marched to a different drummer en route to his position, starting with a master of divinity degree and learning journalism on the job. It may have paid off, as the former religion scholar marches on to number one in both Nielsen and Arbitron on all newscasts.

WTSP-TV is one of the stations that actually cut back on news programming. In the summer of 1988, the St. Petersburg outlet reduced its one-hour newscast at 6 P.M. to a half hour, after a one-year tryout for the longer show. But as a result, Ken Middleton came up with a product far more journalistically exciting than just another local newscast. He gave the half hour left open by the canceled newscast to a special projects unit producing instant news specials, news magazines, documentaries, sports, and weather programs. And by so doing, WTSP-TV is generating a positive, logical direction for local television news. Instead of providing a formula newscast, the station is filling a real need with flexible and exciting programming that gives one-subject depth rather than a TV news game of hopscotch. One show may be devoted to a breaking story such as the pope's visit to Miami, another to the opening of a performing arts center, and, if Hurricane Charlie is hovering off the coast, that's where the priority is. For a change of pace, an artsy half-hour combination travelogue and history lesson by helicopter up the entire west coast of Florida from Pensacola to the Keys may fill still another half hour.

CBS's WCCO-TV, which is number one in Minneapolis, gets a jump on the competition with a unique, innovative show similar to Middleton's. "Newsday" hits the air at 4:30 P.M., with a completely local, live-from-anywhere blockbuster with audience call-in capability. Reid Johnson says it is a back-of-the-book type of news program, heavy in departmentalized features, with one major story a day with guests—a local version of "Nightline."

Survival of the Fittest

The most ominous programming prediction comes from KING-TV in Seattle. Donald Varyu sees a dramatic dropoff in news product from the weaker stations in most markets. "You will see one of the three affiliate stations moving from a realistic local news presence to just a figurehead local news presence," says Varyu, a Chicago and Seattle news veteran. "They will have an evening newscast and a nighttime newscast, but they will fill the majority of their . . . programming time with syndicated material or something that will turn a profit." Varyu also forecasts a general shift in emphasis away from late afternoon.

News executive J. Spencer Kinard considers the direction toward midday and morning news as givens and sees trouble ahead in late afternoon and early evening. Kinard's station, KSL in Salt Lake City, is among those whittling away at the network's morning program. Kinard is negotiating for a full hour of local news between 8 A.M. and 9 A.M., when KSL normally carries the network news show.

EARLY MORNING

The situation in Salt Lake City reflects the fact that the networks are playing in a narrowing field in the early morning as well as the late afternoon time periods. Most local news executives surveyed have already expanded in the morning or have specific plans to do so. They are part of a national trend among more than 80 percent of all network affiliates that now program news from 6 A.M. to 7 A.M. More than half of them devote the entire hour to news. Two hundred and forty affiliates program a half hour of news, while 267 offer one hour, an increase of 60 over a two-year period.

KRON in San Francisco and KGW in Portland both have half hours at 6:30 A.M. KING in Seattle is expanding its present half hour at 6:30 A.M. to an hour, and competitor KOMO has an even bigger appetite. James L. Boyer anticipates expanding his 15-minute cast at 6:30 A.M. to at least an hour. As news director of WWL-TV in New Orleans, his previous post, Boyer took his early morning show to an hour, then 90 minutes.

In Los Angeles, all three network affiliates are already flying the dawn patrol, joining the national trend.

NBC affiliate KCRA-TV in Sacramento, California, is a dramatic example of how far the morning trend can go. The station is adding local newscasts Monday through Friday from 2 A.M. to 5:30 A.M. On the weekend, KCRA-TV is preempting the network's programming for news Saturday and Sunday mornings. Although NBC had to scamper about to place its Saturday morning children's lineup on other area stations, the network is not coming down too hard on the affiliate. In 1980, when KCRA started a weekday 5:30 A.M. to 7 A.M. newscast leading into "Today," it resulted in a big boost in ratings for the network show.

Demographics and money are the primary reasons why newscasts are moving into the early morning. With a substantial increase in the number of working wives, both spouses are getting up earlier. Staffing early morning shows is cheaper because it takes fewer people to put together the newscast. And, thanks to fully automated **robotic cameras** and TelePrompTers that can be run by anchors, the technical support needed is minimal. The high tech dovetails with a penurious but unfortunate tendency to rerun on the early bird shows news packages taped from newscasts aired the previous night. For all these reasons, the temptation is great to invest in an audience that could be hooked in the morning for the rest of the day. In many ways, it's like old-fashioned radio. Viewers are a lot less comatose at that hour, so the early morning newscasts tend to rely less on the visual. Early risers don't have to actually *see* the weather and traffic reports and, if the pictures are fresh or exciting, there's usually time to get to the TV set to see them. Another major factor for local stations in the West and Midwest is that the network morning newscasts, which are videotaped replays, go on the air in their zones hours later and are often outdated by then.

Although the morning expansion is usually not a first-class production, it still costs some money and, as local TV news today is profit-oriented, management would rather not invest new funds in it. As a result, stations find themselves increasingly tempted to tap the original moneymaker, the 11 P.M. news, an ailing institution in many parts of America.

THE VANISHING LATE AUDIENCE

TIO, in conjunction with the Simmons Market Research Bureau, reports that the number of adults watching the late evening news is down 8 percent over a five-year period. The biggest drop was among women, traditionally the market target for that period. The loss of women viewers was nearly double that of men. Research specialist Frank Magid has reported an even more catastrophic decline of 40 percent in some markets over only 18 months.

The numbers make sense. People who get up early to work or play go to bed early. There is also so much news on during the day that people tend to feel overloaded by 11 o'clock. In addition, the quality of the half-hour late night newscast is down. With more than half the newscast gobbled up by commercials, sports, weather, and entertainment reports, the hard news content tends to get short shrift. In certain markets the day's budget of international, national, and local news is crammed into 10 or 11 minutes. And with so many local stations bent on **action news**, that breakneck pace designed to keep viewers tuned in, stories without visuals or "live" possibilities are discouraged. Because the name of the game is a high **story count**, the viewer often faces fragmented news content. The longer 10 o'clock newscasts, with more complete stories, are proving far more satisfying in many markets.

West Coast news directors, in particular, label the 11 o'clock news an endangered species. Midwest markets offering the earlier 10 P.M. newscast and East Coast cities with a later evening lifestyle feel safer.

"It's history," is Harry Fuller's grim prediction for the traditional 11 P.M. newscast in San Francisco. The news director blames changing lifestyles: both husbands and wives, for example, getting up early to work, plunging into a crowded day, and going to bed early. According to Fuller, the shrinking trend has already started on the West Coast— in San Diego, Los Angeles, and San Francisco. In his city, "on a good night the three network affiliates will get a total rating of 19 to 21," compared to the 25 or 30 KGO alone collected for its 11 o'clock newscast just two years ago. Fuller agrees the future is in spread-out rather than concentrated news programming. Yet he has one of the few late afternoon two-hour blocks in the country.

Boyer also has 11 o'clock news woes, but sees a different solution for Seattle. While he doesn't plan to cancel the late news in the immediate future, KOMO more likely will cut drastically the resources that are put into it. "It may become the least important of all my newscasts and I may take the equivalent number of resources that I'm now putting in my 11 and put them at 6 A.M."

Actually, even more dramatic changes may be scheduled for the West Coast. TV network affiliates there are petitioning their home offices for a major change in prime time. They want network prime time shifted from the 8 P.M. to 11 P.M. block to a new one starting at 7 P.M. and ending at 10 P.M. Their reason is the big drops in late night viewing on the West Coast. If network brass were to go along with the change, it would play havoc with late night newscasting. The network stations would, of course, shift their 11 P.M. newscasts to 10 P.M., a period until now dominated by independent stations. The stations had wanted the change to take place as early as fall 1990, but NBC insists it needs more time to study the plan. As a result, the shift is put off to 1991 at the earliest.

One of those independents, KCAL-TV in Los Angeles, figures on getting the jump on all the competition with the unprecedented move of airing *three* full hours of local news, from 8 P.M. to 11 P.M. The station, recently bought by the Walt Disney Company, is making the move in a big way. That includes hiring away anchor mainstay Jerry Dunphy from KABC-TV for $1 million plus a year and beefing up satellite capability as well as studio facilities. For Disney and KCAL, which has been in the news ratings cellar for years, winning is everything. The company is shelling out $25 million in start-up costs for the big show, which is scheduled to debut by spring 1990.

NEWS AT NOON

Like the rest of the country, East Coast stations are expanding into the noon period. CBS affiliates WNEV in Boston and WTVT in Tampa are among those joining a trend that already includes stations surveyed in Minneapolis, San Francisco, Portland, and Seattle.

CHANGES WITHIN THE PROGRAMS

As for the elements within the newscasts, some executives are already reacting to cable's saturation coverage of sports, business, and weather. For example, Fuller has already dropped the sports package from KGO's early afternoon newscast and is pressing his staff to hit the subject from a more creative, more human point of view on the other shows. Fuller and Reagan Ramsey in Portland predict the end of sportscasting as we know it in five years. Ramsey, like Fuller, has had it with the predictable and repetitious rundown of scores and game highlights. As far as KGW's audience is concerned, "People sports is where it's at," says Ramsey.

All-weather cable channels will also have an impact on the use of meteorologists on local shows. In Los Angeles, for example, all four independent stations have dropped weather personalities, giving the chore to the anchors.

MEANWHILE, OVERSEAS . . .

Exchanging newscasts internationally is as easy as punching up a satellite feed. Traveling to other countries, observing news operations, and trading ideas is convenient enough to take the mystery out of what is happening oceans away. As a result, and despite cultural and technological differences, distinct traces of U.S. viewing patterns are appearing around the globe.

However, news programming abroad also has some unique aspects, and one of them is easier accessibility. The scheduling is such that, at any time of day, one of the channels is carrying a news update. Another programming plus is the structuring of channels. Nations such as Japan, Italy, Sweden, France, and Great Britain have channels designated exclusively for local, regionalized news. Viewers don't have to worry about tuning in for local news and getting an international rundown instead. Structuring like this also means generally good relations between local and network.

Japan, like many of the countries visited, struggles to maintain its own identity and yet keep up with American programming techniques. The Tokyo Broadcasting System does a network news program nightly from 6:30 P.M. to 7:20 P.M. The odd concluding time, typical of much Japanese scheduling, was designed to coax viewers into the next half-hour block and discourage channel hopping. However, TBS is considering changing the end time to 7 P.M., just as it is in the United States.

All the network's 25 affiliates do a half-hour local newscast preceding the network show. Unlike that in the United States, however, Japan's program schedule is less competitive on the local level. In Fukishima, Takeo Takahashi starts his local newscast at exactly 6 P.M. Not that Takahashi wouldn't consider getting the jump on the competition by starting a minute or two ahead of them in the near future. TUF-TV does only a three-minute newscast at 11 P.M. because right now that is "all TUF can afford to put on the air."

Long, late night network news is not unusual abroad. For example, TV Asahi airs a flexible national cast that runs from an hour to an hour and a half.

NHK scheduling would make any U.S. TV prime time programmer wince. The government-controlled network provides a network schedule that includes a 20-minute newscast at 7:30 A.M., five minutes at 12:15 P.M., a showcase half-hour cast at 6:30 P.M. and a final 15 minutes at 8:45 P.M. NHK's local newscasts could hit at any time, depending on the affiliate. In Tokyo, "News Center 9," the local evening news, broadcasts from 9 P.M. to 9:40 P.M.

In Italy, you punch up the TV and take your choice of news from 7:30 A.M. until well after midnight. News programming is far more prolific than in the United States, but it's as confusing as the politics. TG-1 produces a "Today" type show from 7:30 A.M. to 9:30 A.M., a 5-minute newscast at noon, 30 minutes at 1:30 P.M., five minutes at 6 P.M., 30 minutes at 8 P.M., 10 minutes at 9:40 P.M., and another half hour at midnight. Socialist TG-2, which specializes in nonpolitical sports, schedules most of its programming late in the day, capping it with a newscast described by executives as running "30 to 40 minutes" at midnight. If Italians want regionalized news, they go to Channel TG-3. It does three regional newscasts in the evening.

Sweden has two networks: Channel 1 for national news and Channel 2 for regionalized news. Network news is on for five minutes at 3 P.M. and 5 P.M., 10 minutes at 6 P.M., and 45 minutes at 9:30 P.M. Local news hits the air in 15-minute packages at 7:15 P.M. Monday through Thursday and a half hour on Friday nights.

Overseas Privatizing

Both Italy and France are experiencing an incursion of private TV station ownership and, at some point, this trend could have a major effect on news programming. In both countries, industrial magnate, entrepreneur, politician, and former showman Silvio Berlusconi of Milan is the new proprietor on the block. And in both countries, his is Channel 5. However, journalists in France and Italy point out that, while Berlusconi has experimented with news, he has found it unprofitable, and it is unlikely Channel 5 will continue news programming. That doesn't mean it won't occur to the enterprising magnate to take a long, hard look at America's commercially successful TV news and imitate accordingly.

Like Italy, but unlike the United States, France feeds its audience's news appetite throughout the day. For example, Antenne 2 produces newscasts starting at 6:45 A.M. That first show, described by Elie Vannier as "a cross between the 'Today' show and 'Good Morning America,'" runs an hour and 45 minutes. Between 9 A.M. and 6 P.M. there are eight newscasts varying in length from two and a half to three minutes. In New York or Washington, these would be headlines. But in France, despite the telescoped length, Antenne 2 inserts videotape or a live shot from breaking news stories. At midday, Antenne 2 has a 40-minute newscast, and from 8 P.M. to 8:30 P.M., its major prime time newscast. The last show of the day, a half-hour combination of news and ABC's "Nightline" format, hits at 11:30 P.M.

To make sure the French borrow equally from all three U.S. networks, Antenne 2 offers a **magazine show** described by Vannier as in the style of CBS's "60 Minutes."

In Great Britain, too, news junkies can get a fix any time of the day, with the added bonus of having it delivered in impeccable English. Publicly owned BBC 1 has a solid four and a half hours of news and current events a day, including an hour and a half in the morning, hourly summaries during the day, 30 minutes at 1 P.M., 35 minutes at 6 P.M., and 25 minutes at 9 P.M. BBC 2 is limited to a program called "Newsnight," which airs late evening five times a week.

Liz Forgan, deputy programme controller, is proud of Channel 4's only newscast, a 50-minute show that downplays breaking news and airs at 7 P.M. The innovative, more talk than action, depth-conscious program is considered by other British broadcasters experimental and with a questionable lifespan. Americans will see in it traces of such exemplary U.S. programs as the "MacNeil-Lehrer Newshour" and "World Monitor," the half-hour magazine program produced by the *Christian Science Monitor*.

As for local and regional news coverage, programming is equally as intense as the BBC and ITV's network schedule. Just ask Thames's Barrie Sales. The independent outlet in London is responsible for a four-minute newscast at 9:25 A.M., a two and a half minute cast at 11:25 A.M., five minutes prior to 1 P.M., another two and a half minutes at 3:25 P.M., the major half-hour local blockbuster at 6 P.M. (following the ITV network newscast), and another short wrap-up at 10:30 P.M.

Sales's counterpart in Birmingham, England, Laurence Upshon of Independent Central Television News, has a similar workload—half-hour newscasts on weekdays plus 52 news bulletins a week, varying in length from one to five minutes.

His local competitor, the BBC's Rick Thompson, produces 35 news inserts a month, from breakfast through the evening, plus all the radio newscasts for his Midlands operation.

The international news world is a big one. It projects an impressive volume of news and does so with the intriguing features of flexibility and choice.

PROJECTION

Local U.S. TV news programming is heading for a far more satisfying schedule. And, at the rate it is progressing, it could very well take on some of the characteristics of the foreign operations. Local news will be on the air morning, noon, late afternoon, and night in the immediate future. That big block of local news in the afternoon, however, will be broken up in order to make the money and personnel available during the other time periods. While the hour-long show is still tempting to some news producers, most local newscasts will soon shrink to a half hour. In addition, as more people watch news in the early morning as well as midday, it is reasonable to expect that they might watch between these times as well. This will lead, within five years, to additional, short bursts of news—perhaps even on an hourly basis, throughout the day and eventually around the clock. One station, KMOX-TV in St. Louis, has already begun doing this, with 35-second hourly updates every hour, every day. It also presents weather and time on the half hour.

As long as there are viewers, there will be advertisers insuring the local news operation's role as a profit center. High tech will lead to a reduction in the number of technical people a station needs, thus making it increasingly easier to make those profits. On the plus side, technology will also give the shorter newscasts the ability to patch in field reports from any place any time. Guardians of the purse strings will take a hard look as well at any programming that loses money, and the results will be ruthless. The 11 o'clock newscasts on both coasts will shrink or disappear, mainly because of changing lifestyles.

The content of programs that remain on the air will undergo dramatic changes thanks to competition from cable and other TV stations. Count on streamlined and more humanized sports, weather, and business news.

Technology will continue having an impact on U.S. as well as foreign programming. As we are quickly learning, the marvels of the 21st century are already here, which means more satellites over more countries providing even greater variety and availability of news programming.

8

---□---

ANCHOR . . .
OR ALBATROSS?

Anchors help sell the newscast, but the danger is cosmetic journalism and an erosion of show and shop control. Paying the star with megabucks and power can mean giving away the farm as well as megaheadaches for personnel and product.

French journalists may have the right idea. In the fall of 1988, when reporters at the government's Antenne 2 discovered that anchor Christine Ockrent had been hired for an unprecedented $200,000 a year, they went on strike. In France, broadcast journalism is no frivolous matter. The walkout spread from the state-run networks to the commercial stations, knocking half of them off the air. It was also considered the first real political threat to Premier Michel Rocard's government.

While the strike was brief and resulted in no substantial difference in policy, the Ockrent affair did stir up a fuss. As is the case with almost any news or gossip about TV in France, it set off an avalanche of letters, newspaper articles, and political cartoons. Ockrent's super popular 8 P.M. newscast—limited to a handful of scab reporters and wire service copy—suffered temporarily. And the rhetoric gave many affected persons a chance to vent feelings. For example, striker Silvie Marion, a reporter for Antenne 2 for 20 years, summed up her bitterness for the press: "It would take me 12 years to make what she will make in one year." And even French president François Mitterrand got into the act by tipping toward the anchor, with the comment, "Talent and work must be compensated."

Claude Contamine, president of Antenne 2, had coaxed Ockrent from the private network by doubling her salary, an offer that predecessor Elie Vannier probably would not have made. Vannier, a no-nonsense journalist, has a 20-year career in journalism. During his reign at Antenne 2, he fought to keep anchor salaries and influence in line. In 1988, he fired his top male anchor because the newscaster insisted on stating his opinions on the air.

What Vannier was up against, and what many local U.S. news directors fight today, is upper management's willingness to pay anchors runaway salaries and give them significant control because of their popularity and profitable ratings. It is the same syndrome that has viewers referring to it as the Tom Brokaw show instead of "NBC Nightly News."

Brokaw, and Dan Rather, the $2.5 million per year man, are managing editors as well as anchors. Bryant Gumbel, the NBC news department's "Today" anchor, has a contract that pays him $7 million over three years. While the former sportscaster is not managing editor of "Today," he does wield significant power over the program. Staffers are still smarting from Gumbel's highly publicized "secret" memo to executive producer Marty Ryan that strongly suggested major changes in on-camera and off-camera talent.

Power, as well as an increase in salary to some $1.3 million a year, was a factor in Connie Chung's move from NBC News back to CBS in 1989. According to NBC officials, she left because the network wouldn't give her complete editorial control over a new magazine program, the right to name her own executive producer, assurance that she would be its sole anchor, and significantly more money. At CBS Chung got a $400,000 raise and replaced "West 57th Street" with "Saturday Night with Connie Chung." NBC then quickly signed up Mary Alice Williams to help fill the vacuum left by Chung. At CNN, Williams was a vice president, New York bureau chief, and co-anchor of two news-casts. Her salary at NBC, $400,000, is twice what she earned at CNN. Significantly, the peacock network executives also indicated they would probably hire three "seasoned reporters" with the $900,000 they had been paying Chung. The flurry of musical anchor chairs also danced Diane Sawyer from CBS's "60 Minutes" to co-anchor a new show with Sam Donaldson at ABC for $1.4 million a year.

In local TV news, it is no longer unusual for a top-market news or sports anchor to pass the million dollar a year mark and carry a title that gives him or her authority. According to an RTNDA *Communicator* survey, the typical anchor salary at a high-paying station is $1,105,000, the highest reported being $1.56 million. A writer, pro-ducer, or news director who makes a fraction of those seven-figure salaries is going to think twice before disputing the anchor's choice of a lead story.

The newly molded hyphenate of anchor–star–managing editor is having its impact on the role of the reporter as well. Former NBC reporter Lloyd Dobbins wrote in the April 16, 1988, *TV Guide* about the trend of giving the main course to the anchor and the leftovers to the correspondent in the field. Dobbins referred specifically to Brokaw's increasing tendency to present the important information in the story, leaving only the background to the reporter in the field. "It's the current trend at the networks to make their anchors ever-rising stars of intelligence and significance."

Dobbins also reflected back to the NBC of the 1960s and the "Huntley-Brinkley Report," which ended with Brinkley's "Good night, Chet" and Huntley's "Good night, David, and good night for NBC News." The implication was that the anchors represented the many other people who made the program possible. Compare that with the current NBC sign-off of "That's 'Nightly News' for this Wednesday night. I'm Tom Brokaw. I'll see you tomorrow night." The impli-cation is that "Nightly News" is a one-person show. At CBS, of course, it is more than implication. The name, "Saturday Night with Connie Chung," clearly spells out the dominance of the star.

THE ANCHOR BEAUTY TRAP

Jo Moring isn't mourning the decline of the star correspondent. She says, "In the purest sense, the story and the people to whom it is hap-pening are the most important." Moring also feels a good reporter in the field gathers it all, sends back all the facts—and so what if that reporter happens "to be fat, bald and not a particularly vocal person

. . . and Tom Brokaw, who is not fat, bald, and unattractive happens to voice the story. Is the story any less valid?"

Viewers don't seem to care. They have no trouble accepting CBS anchor Charles Kuralt's chubbiness, former NBC correspondent Edwin Newman's thinning hair, KCBS-TV anchor Bree Walker's deformed hands, or any of the litany of blemishes heaped on KRBK-TV anchor/ news director Christine Craft by her former management. The Craft case symbolizes the cosmetic approach to anchor hiring and firing in America. Craft is now a successful broadcast journalist at KRBK in the country's 20th largest market, Sacramento. But in the early 1980s her world came apart. The manner in which KMBC in Kansas City fired her is now legend. According to Craft, in the book she wrote about the experience, *Too Old, Too Ugly, And Not Deferential to Men*, this is what management told her:

> Christine, our viewer research results are in and they are really devastating. The people of Kansas City don't like watching you anchor the news because you are too old, too unattractive and you are not sufficiently deferential to men. We know it's silly, but you just don't hide your intelligence to make the guys look smarter. Apparently the people of Kansas City are even more provincial than we had thought. They don't like the fact that you know the difference between the American and the National Leagues! We've decided to remove you from your anchor chair effective immediately. You can stay on and continue your reporting and earn the rest of your contracted salary, but just remember that when the people of Kansas City see your face, they turn the dial.

Craft sued Metromedia and its station for fraud and sex discrimination in 1983 and won, but lost the appeal.

Among the many ironies is that Craft is attractive and presentable as well as talented and experienced. Craft points out another irony in her book. "Fawn Hall, the document shredder who testified under a grant of limited immunity that she never 'asked questions,' is courted by local television stations for a job that requires being very good at asking questions." And this is occurring "as veteran newspeople like Fred Graham and Marlene Sanders are cut loose at CBS because the bottom-line people can't see the value of seasoned journalists." Reporters Graham and Sanders were let go for "economic" reasons.

The case of Catherine Crier at Cable News Network is another example of cosmetics getting the edge on experience. In October 1989, the extremely attractive Crier was teamed up with Bernard Shaw to anchor CNN's first prime-time newscast. Crier, a former state district court judge in Dallas, never worked in TV news before.

On the other hand, some journalists feel a reverse prejudice. Deborah Norville, who replaced Jane Pauley on "Today," maintains she suffers from her good looks. Norville, in the November 6, 1989, *Newsweek,* told reporter Jonathan Alter: "The blue eyes, the blond hair, the gender have been a handicap. They may have gotten me noticed, but they would have gotten me thrown out three times as fast if I hadn't busted my butt." Norville, a *summa cum laude* college graduate with 12 years experience in TV news, claims people expect her to be unqualified because of her attractiveness. The journalist insists the cosmetics have forced her to work longer hours and do more stories.

Like so many aspects of show biz and TV news biz, the decision to go with a pretty face is often based on assumption rather than reality. While TV executives tend to look at the broken nose or gray hair as taboo, viewers may not see it that way. How journalists are promoted often plays a role in their acceptance.

THE PROMOTION TRAP

Promotion does have its share of paradoxes. Slick, intensified campaigns that make stars out of anchors have become time-consuming, ego-building monsters. The scenario is likely to be scripted by outside professional advertising agencies, and the result is a crowded schedule of still-taking, promo-shooting, award-getting, and community-involving activity. That doesn't leave much time for celebrity anchors to report or write.

Another major by-product of the promotion trap is its impact on the credibility of the newscasters and their newscasts. Stations lean toward commercial style **promos** more suitable for selling soap than news. In Los Angeles, one promo has the morning news anchor and his team step into the bedroom of a sleeping couple to let them know the show has been expanded and is going on earlier. Elsewhere, anchors have been dressed in bizarre costumes, covered staged stories, and performed side by side with actors for the sake of promotion.

An unfortunate side effect of these campaigns is that people then tend to watch and relate to the star anchor rather than the story or newscast. Some news directors recognize this and have learned how to survive. James L. Boyer, for example, admits, "That is how the public chooses and, until that changes, it will never change. The person who is the human contact between the television station and the consumer of the news, that person becomes very important." As far as the KOMO news director is concerned, it doesn't matter how good his reporters are, how good his stories are. "If the public doesn't like my anchor, they aren't going to watch my broadcast."

The star syndrome leads to certain inevitable results. In March 1988, for example, *Esquire* magazine sported a sexy display of Meredith Vieira's legs. *Vanity Fair* had a sultry shot of Diane Sawyer on the cover. And *Harper's Bazaar* did a spread on Maria Shriver's makeup routine.

With sexy, superstar, overpromoted anchors, the personality dominates the news event. At news events, it is not uncommon for crowds of autograph seekers to surge toward the more recognizable anchor, completely ignoring the newsmakers themselves.

ANCHOR CONTROL

The one negative trend in local TV news Ernie Schultz admits to increasing anchor dominance. "Eight-hundred-pound gorillas," he calls them, who "jerk people around . . . jerk producers . . . jerk news directors around." The RTNDA executive zeroes in on oldtimers "who were never qualified to do the news," who draw "inordinate salaries," and who "have acquired great power that has nothing to do with ability or experience or anything else." While Schultz agrees that anchors are the biggest single factor deciding which station is watched, he still wants more emphasis on the reporter. Schultz wants to develop a corps of reporters so strong they drive the anchor people back into the background. As for excessive salaries, he wants that money to go into the product, not the presenter.

Giving an anchor editorial control is bad enough for the team psyche, but the problem is greatly exacerbated by the fact that anchors, like news directors, have short lifespans at TV stations. As a result, the anchor–managing editor may not be familiar with the geography, demographics, or past or recent events in the local area.

A case in point involved the sensational Patty Hearst case of the 1970s, including her bizarre kidnapping, involvement in an armed bank robbery, a bloody shootout between the Symbionese Liberation Army and the police in Los Angeles on live TV, and a nationwide search for the millionaire heiress. Hearst's capture broke just prior to our KABC-TV late afternoon newscast. The author, as acting news director, gave the story top priority, ordering up video lines from the capture site, San Francisco, and clearing the way to devoting most of the newscast to a story that excited Los Angeles, the state, and the country.

The newly imported anchor–managing editor didn't see it that way. "What's all the fuss about?" he whined as producers and writers scurried about, trying to beat the competition on a major breaking story. In fact, he insisted, the Hearst story should be given the same treatment as any of the routine crime stories in the cast. So in the valuable minutes before air time, there ensued a shouting match bordering on violence before the entire newsroom. The author wasted valuable time explaining the importance of an obvious story and giving a short course in rudimentary journalism. The only solution was to override the anchor's managing editor status and give a direct order, at the risk of inciting management reprisal. It worked and Channel 7 beat the competition, but the losses, in terms of morale, harmony, and team spirit, were heavy.

IMPACT ON THE CORRESPONDENT

As for the networks, Robert Mulholland puts it this way: "The whole news department is the supporting cast for the star performance by the anchor person." He, too, sees the impact of the anchor on the status of the correspondent. A "deculting," he calls it. "The worst job in the world these days is probably that of a network news correspondent. You've got financial pressures on one hand and you've got the anchormen who are running the news division on the other. And you're in the middle trying to deal with both of them."

In addition, correspondents often live out of a suitcase and don't see their families for months on end—a life like this has a way of taking all the romance out of being a network news correspondent. Further, network correspondents are joining local reporters in the high-tech approach to covering the news. As camera and recording equipment get more compact, so does the technical crew. News crews used to be made up of reporter, camera operator, assistant camera operator, light person and sound person. Then there were three: reporter, camera operator, and sound operator. Then, in many markets, two: reporter and camera operator. Soon, as is already the case in the smallest of stations, there will be one: reporter–photojournalist.

For management, this may be a great way to save money, but for the reporter–photojournalist it creates several problems. There is a conflict between concentrating on the technical shooting procedure and paying attention to the editorial thrust of the story. In addition, because shooting a news story takes all the attention of a camera operator, valid concerns arise about the operator's personal safety. Peering into the **viewfinder** and worrying about the shots can create tunnel vision, which means that much of what's happening around the operator is blocked out. With no one left to watch out for approaching problems or actually guide him or her around obstacles, the photojournalist is left in physical danger. It's easy for dedicated camera operators to get carried away in their efforts to get that perfect picture—they will often shoot out of helicopter doors without seat belts, for example. Physically holding onto the shooters and protecting them in a number of ways are tasks that other crew members take on automatically.

CONTROLLING THE ANCHOR

Many news department heads are concerned about anchor dominance and are making serious efforts to deflate the star status. Larry Price has a hard-nosed approach to high-paid anchors in Minneapolis and predicts stations will soon start refusing to pay runaway salaries. KSTP-TV also gets more from its anchors. They are expected to participate in the show, be the newsroom's best reporters, and go out on assignment on a regular basis. The ABC affiliate's anchors must excel in live situations and emergencies, have good communication skills, *and* look good.

Herbert Dudnick is more concerned about anchor control than inflated salaries. "The biggest mistake they ever made was to give Brokaw, Jennings, and Rather as much say in the show. Either you have a producer or you don't have a producer. And that's it." As for KRON-TV, anchors do not run the news operation. Dudnick does.

Harry Fuller claims KGO, too, has become tough and is getting rid of the overpaid star concept. Fuller, who recently fired a $500,000-per-year-anchor, claims it's going to be at least a decade before anybody at KGO approaches that kind of money again. He acknowledges the importance of personality, but insists that people are beginning to realize the anchor is just one of several ingredients.

Fuller traces the anchor monster to the networks' arrogance and aloofness. He says the networks are run by people in New York who have no idea how the real world runs because they've all worked in the Big Apple too long. "When is the last time one of those guys drove up to a shopping center and went to the Safeway and shopped for food?" Mike Wallace and his colleagues working on "60 Minutes" are the "seven millionaires," not real people.

The more serious problem, Fuller agrees, is control. "When you cross the line and make him a managing editor like they did with Dan Rather, that's when the real problem starts." In this area, he claims, local stations are light years ahead of the networks. "The local station that's run right knows it can't give editorial power to somebody who's an anchor. You don't want the star directing the movie."

Nor does Reagan Ramsey mince words. He calls stations that automatically capitulate to anchors "stupid." But the Portland executive is optimistic and thinks the affiliates are waking up and refusing to pay the big bucks. The peak, according to Ramsey, was about 1985, but today many stations realize there are a lot of people who can do these jobs, and losing an anchor doesn't mean going off the air. At KGW, "Anchors don't dominate any more than a reporter, the assignment editor, or the ENG editor does. They're a key part of our organization, a key part of our newscast, but not the whole newscast." Ramsey's anchors are former reporters. They write, report, and do live stories as well as live newscasts in the field. The route to the anchor desk at the NBC affiliate is not a beauty pageant. Ramsey's criteria are whether the person is a good communicator, a good writer, and a good journalist.

Portland is in sharp contrast to Los Angeles, which seems to work overtime to retain the Tinseltown nickname. Among its anchors are a former Miss America, a secretary, a traffic reporter, and a sportscaster.

THE AGENT

Many credit the 10 percenters for the crazily spiraling news salaries. It was an agent who pulverized the payment formula by negotiating six figures each for Huntley and Brinkley back in their heyday. Another big breakthrough came years later, with Barbara Walters' contract— the first million-dollar salary to be negotiated for an anchor.

Since then, talent agents have played a significant role in escalating salaries in the networks as well as the stations. Yet, they are often necessary. In the big markets, anchors do bring in big money and deserve a fair shake. And in the very small ones, they often do not make a living wage. It is not unusual for an anchor–reporter–producer–photojournalist–editor in the lesser markets to bring in as little as $12,000 a year, which doesn't even pay for TV-presentable clothes or hair style.

But the big guns have spoiled it for the others and some stations refuse to deal with any agents. KGW is among those stations. There, Reagan Ramsey tells prospective anchors, "If you want a job here it's between you and me." Not that talent isn't encouraged to have a lawyer go over the contract. To news director Ramsey, the agent is "an unnecessary parasitic element of the business."

In Seattle, Donald Varyu will talk to agents, but blames them for the anchor dominance problem. He says they have negotiated power for their clients by demanding titles or influence in exchange for less money. Varyu admits he has anchors who make more money than he does, but none of them carries the title of managing editor or supervisor. That's something you don't have to give away, says Varyu, "but a lot of people have, in the name of allegedly saving money, because it soothes some egos. That's what creates the egos." However, Varyu doesn't have an anchor problem at KING-TV. He runs the show.

MIDDLE GROUND?

James L. Boyer seems to try to follow a middle road, but the forks along the way are tricky. On the one hand, he says, KOMO-TV producers are in charge of the broadcast but, on the other, he admits anchors have all the input they want. Boyer's policy is that, if anchors "see something they don't like, I expect them to say, 'I don't like it.' There's no way a producer can make an anchor read something that the anchor refuses to read." In other words, the anchor has the final say as to what goes on the air. Boyer recognizes that producers are responsible for newscasts, but, "At the same time, it is the anchor's mouth. It is the anchor's image, anchor's face and anchor's voice."

Yet an anchor's popularity is not the only criterion at the Seattle affiliate. Boyer reflected on a recent year-long talent search that he thought was finally over with one particular person whose market testing was sensational. But, according to Boyer, there was no way the anchor was going to get along with the staff. Rather than have a newsroom in turmoil, Boyer chose not to hire him.

The runaway salary is a problem for the industry, admits the news director, who says he pays anchors competitively. Certain talent, maintains Boyer, is worth the money. His position on Walters' seven-figure breakthrough at his parent network: "ABC wouldn't be paying her a million dollars unless she was worth a million dollars to them." Yet, Boyer bemoans control problems elsewhere, at stations he says relinquish control to the anchors out of fear. He claims those presenters do whatever they want because general managers and news directors, who together don't make as much as the anchors, are afraid for their jobs.

J. Spencer Kinard is in sync with Boyer's credo, contradictions and all. The Utah news executive agrees that most people decide, on the basis of the personality of the anchor rather than on content, which television newscast to watch. Kinard admits his news and weather anchors are paid three to five times more than any other employee, but their salaries are not as runaway as in other parts of the country.

The CBS affiliate news vice president sees a slowing in this salary trend, but he feels the reason for the slowing is a tighter newsroom economy. Because Kinard insists that anchors not be hired at yesterday's exorbitant salaries, pay raises have come down substantially. However, he is quick to admit that, if ratings numbers slip and replacing an anchor will cost a lot, the station will be forced to do so.

KSL is also not ready to rattle the gorilla cage. On the one hand, Kinard claims that anchors are not managing editors and don't control the newscasts, that the producer is in charge of the show. On the other hand, though, Kinard admits that if anchors want to decide what the lead will be, overruling the producer, "we coddle them." Anchors, confesses Kinard, are "important enough to us that we don't want to make them mad so they leave."

THE GENERAL MANAGER'S INFLUENCE

One of the major problems in keeping anchors under control is the station's general manager. It's the general manager who hires the anchor star and shells out the big money. Naturally, the boss wants to protect his investment, and the newscaster is well aware of it. A general manager enjoys showing off his celebrity and the two of them spend a lot of time together, eating, drinking, playing, *and* talking. That's a tough situation for the news director who, according to the organization chart, is supposed to be in charge of the talent.

Kinard is one of the few news executives to acknowledge the power of the general manager in these delicate relationships. Too many anchors are on a power trip "because they know they've got a direct line to the general manager and the general manager is going to keep them happy."

The problem, Kinard insists, is the business itself. "Because of the salaries, we've created these egotistical monsters who think they are show business stars . . . try to act like it, and dominate every decision that's made only to their interest, not to the public's."

There's one news director who doesn't see gorillas behind the anchor desk. Ian Marquand gets along just fine with his other three Helena, Montana, staffers, all of whom are the ultimate hyphenates, producers–reporters–writers–editors–photographers–anchors. Personality problems? Nobody's got time for them. Fights over controlling the newscast? They'd be punching at themselves. Runaway salaries? Marquand is more concerned about getting them $2000 raises.

ANCHORS OVERSEAS

At least two countries, Japan and Great Britain, are joining France in tackling anchor problems, a relatively new headache for them.

In Japanese broadcast journalism, all employees have had the same salary regardless of job titles *until now*, reports commercial network TBS. Although there are still no $2.5 million anchors, Junji Kitadai sees such a trend developing in Japan. He points to a dramatic shift toward hiring newscasters on a contract basis in order to pay them more than the other employees. For Japan, this is a major cultural and professional change. The country prides itself on creating a family environment in its companies, and it is traditional for workers to stay with the same employers until retirement or death. The security is incredible and the companies pay for all the needs of the employees and their families. But now, Kitadai says, TBS gives contracts to popular, highly rated news presenters. "So, in that sense, they are stars."

Commercial TV Asahi's Takehiro Irie flatly admits his network sees profits in its presenters. Unlike the competition, TV Asahi's anchor has no news background. The news program director admits that Asahi's top anchor, who doesn't write, came from an entertainment background, specializing in master-of-ceremonies jobs.

At more conservative, government-controlled NHK, the anchor is "one of us," says producer Kenichiro Iwamoto. "He represents the program and he expresses his opinion as to how the program should be made and we respect that, but we can say all kinds of things to him as well." The NHK anchor doesn't carry the managing editor title. In Japan, all members of the team are expected to pitch in.

NHK anchor Yoshihiko Kubota contributes directly to the script, changing copy he doesn't like, but he calls it a forward step from the time news was read by announcers. Kubota earned his anchor position. Like many colleagues, he started at NHK with no journalism training or experience. NHK taught Kubota reporting, a job he had for 20 years, until 1985, when he was promoted to newscaster.

In Great Britain, BBC executives admit anchor salary problems. However, Chris Cramer thinks it will be a while before anchors approach the Dan Rather or Diane Sawyer mark. British anchors' salaries range between $72,000 and $180,000. The problem is that the average producer working on national TV news or radio earns only about $32,000, and average reporters get at best some $63,000. These salaries are particularly low compared to those in the United States. Local producers here, according to RTNDA surveys, can earn as much as $75,000; news directors, $161,200; reporters, $447,000; and anchors, above a million dollars.

BBC anchors used to be readers, but, reports Cramer, the move in England is strictly toward journalist–anchors. "We've always had the view that people ought to look personable on television, but the important thing is that anchors now have to be journalists first, presenters second."

In the Midlands, Rick Thompson can see the kinds of anchor problems that plague the United States cropping up in Britain in the near future—ironically, he adds, because the anchors *are* now journalist–anchors. He says the title opens the door to giving them input on the selection of stories, on the way stories are written, and on the way they are presented. However, BBC presenters can make money elsewhere by moonlighting. They are permitted to participate in supermarket openings, sales conferences or corporate video presentations, but they cannot appear in commercials.

Does the BBC really worry about superstars and super salaries? No, insists Thompson, referring to England's "massive traffic jam of superstars asking the BBC for money it can't afford, being turned down, going elsewhere, and disappearing into oblivion." A typical anchor salary in the Midlands is $45,000 a year. Reporters are not that far behind, with a salary of just under $38,000. Producers average $30,500.

The British maintain that cosmetics is not a factor. Both Central and Thames TV executives say their anchors are airworthy if they are authoritative and have tremendous experience as journalists.

One reason BBC's competitors act somewhat more secure is that they pay more. In Scottish TV, for example, anchors get $59,000 and reporters earn above $42,000. Producers do even better than that, taking home approximately $48,500 a year.

BBC employees, like their counterparts in France, demonstrated against the higher salaries in the commercial companies with strikes in 1989. Some 18,000 BBC journalists and technicians walked off their jobs.

While Sweden's cost of living rivals that of Japan's, the TV news salaries don't. In Sweden's Örebro region, anchors get about $26,400. Reporters, depending on the amount of experience they have, make slightly less, and producers collect about half the anchor salary. But unlike in the American pattern, the news director does better than his talent, pulling in about $40,000.

Janne Andersson reports a questionable trend. While one of his anchors, a male, came up through the ranks as a reporter, the other, a female, is cosmetically more appealing. The anchorwoman is a recent university journalism graduate who has been working for some six months. However, Andersson denies she was picked only for looks. Like broadcasters in the United States, those in Sweden also cater to demographics. One of the reasons the recent journalism graduate anchors is her Hungarian heritage. Andersson says there are too many people in Swedish television looking just like Swedes. A million people out of a population of 8 million come from abroad.

Although Swedish anchor salaries may be pocket change to most American broadcasters, the Swedish anchors don't see as much air time. As in Japan, Swedish TV rotates producers and anchors. Anchors on Andersson's TV 2 ABC regionalized newscast are limited to working 11 nights a month.

French Channel 3 in Paris rotates anchors every other week. However, anchor Remi Guidicelli also runs the heavily populated Paris region bureau, produces the daily 20-minute newscast, *and* reports. Channel 3 is the exception. Elsewhere, with Ockrent's salary explosion, France moves toward America's value system, as do other countries throughout the world.

PROJECTION

It is tempting to look at the U.S. star system as simply a reality that has to be accepted by news management and the viewer. After all, whatever is on TV is show biz and, if anchors are good box office and help bring in the bucks, they should share in the profits. There's certainly precedence for letting the star influence the product as well as the other people involved. Those other stars, the ones in the movies, often play a part in choosing and monitoring directors, writers, and other actors.

One problem, however, is that the star system, born in the news explosion of the 1950s and 1960s, escalated out of control. The million-plus salary, added authority, promotion, and star status are having an impact on the traditional role and responsibility of news management. Although in some cases anchors may be bona fide experienced journalists, they rarely have the time to devote to the newscast. Because of the musical chairs syndrome, they may also not be familiar with the market, the station, or its personnel.

The other problem is of greater concern. Journalism traditionally has been about public service, not commercialism. The more emphasis that management and the public put on box office appeal, the more likely TV news credibility will suffer. For years, broadcast journalism was on a roll, with people clearly choosing it as their sole source of news and judging it consistently believable. However, the politically conservative United States of the 1980s created a mood of mistrust toward the media—not the best time or climate for journalistic superficiality.

As for the future, there is hope, but it may be at the expense of the networks. Desperation is the problem. The network news divisions, in an effort to survive, are in near panic competition. The more popular anchors, taking full advantage of the intense fight for the rating point, are making outrageous monetary and power demands and getting them. If anything, the trend could hasten the demise of the big three.

As network anchors demand and get more exposure, the grunts, correspondents, will get less. As a result, even more reporters will trade in their network stripes and be lured to the affiliates. For the stations, that's a plus; they will be getting experienced, mature, knowledgeable reporter–anchor journalists with name value. It might also mean a return on camera of the gray heads, a fitting reward for the discards of the 1980s.

As for runaway salaries and power, some affiliates are already drawing back. Unlike the networks, the stations are clearly on the positive side of the profit-and-loss ledger as well as the ratings book. These highly motivated managements can afford to be far less tolerant about escalating costs.

If stations could work together to deflate anchor salary and control, they could create a new environment. In such an environment, qualified, equitably paid, and recognized anchors, reporters, producers, and writers could more easily pursue their mission: honest journalism.

9

---◼---

FOREIGN INTRIGUE

A Closer Look
at TV News Overseas

Their broadcast journalism tends to mirror ours. Beyond the reflection, which shows our good and bad habits, they are doing some exciting and innovative things of their own "over there."

America is the worldwide leader in TV news, which means that many of our journalistic exports are clearly visible on foreign screens. To a great extent, looking at newscasting overseas today is looking at practices that brought us to where we are in the United States. Seeds have been planted in Europe and the Far East for ratings races, a faster newscast pace, happy talk, cosmetics, superficiality, and sensationalism.

At the same time, other things are happening abroad that could be of value to broadcasting in the United States and those who are about to be part of it. Not that all positive journalistic goods from abroad are necessarily altruistic. For example, several countries, such as Japan, Sweden, Great Britain, Italy, and France, don't waste energy in affiliates versus networks combat. But the reason is that many of these networks are government-controlled, and internecine competition is strongly discouraged.

THE DANGER OF REPRESSION

Government control can also lead to government interference. Although BBC and ITN can be very good, broadcast journalists in the United States do not envy the recurring problems British TV news has had with Margaret Thatcher's government. Nobody seriously contested the prime minister's strong influence on curbing gratuitous sex and violence. However, Great Britain's new Official Secrets Act and measures aimed at combating terrorism in Northern Ireland are eroding the nation's traditional media freedom. The Secrets Act commits the security and intelligence services to permanent silence. Another measure forbids TV and radio statements by members of terrorist organizations, including Sinn Fein, the political wing of the Irish Republican Army. The government has also tried to interfere directly with the broadcast of a Thames documentary, "Death on the Rock." The program, which questioned the shooting of unarmed IRA terrorists, was condemned by the government, which tried to have it banned. An independent inquiry eventually resulted in the program's broadcast, but freedom of information suffered.

In spring 1989, the situation worsened and culminated in a condemnation of the Thatcher government by the International Federation of Journalists. The Federation accused the government of "systematic and extensive" muzzling of the media. According to polls, including one ironically produced by government-controlled BBC, a majority of Britons are seriously concerned about Thatcher's repressive attitude toward electronic and print journalism. The irony is that Britain's dedication to civil and political liberties precedes our Bill of Rights by centuries. Their guarantee dates back to the year 1215 and the signing of the Magna Charta.

Despite political, cultural, and social differences, attitudes and expertise in other countries are well worth noting. The following is a sampling of the duty-free journalistic goodies that rank right up there with the best scotches and perfumes.

1. *The ability and courage to tell a complete story*

Several countries don't understand our love affair with the fractured news story. Coverage in depth is among the things they do better.

NHK is not afraid to run a "big story," anywhere from 10 to 40 minutes, and follow-up is a priority for the Japanese. TV Asahi has a firm policy calling for a special segment of at least five minutes on a single subject. It can be a more detailed treatment of a breaking story, an investigative report, or a thorough look at any problem affecting a large segment of the audience.

Sweden's regionalized newscast earmarks four or five "bigger" stories that warrant more time on the air. Of those, Janne Andersson picks one that will run "two or three minutes," an unlikely scenario for a comparable 15-minute newscast in the United States. The news chief's credo is to report what happened as well as why it happened and what it means for the future. Recent three-minute pieces included the attempt by Stockholm's legitimate theater to get government subsidization and a graft situation involving the building of a hotel in the north. Ingemar Odlander, the producer responsible for Sweden's nightly half-hour national newscast, prides himself on tackling boring subjects, such as the Common Market, and making them come alive with extended treatment.

The BBC's Chris Cramer, who feels cheated by our tightly edited news, devotes more time—an average of about two minutes—to telling stories. Barrie Sales claims an average length of stories on Thames's longer newscasts of three minutes and, if the story is big enough, it will take over the entire show.

Tools for Depth Coverage

Thames's policy is to react to whatever is important in London on that day and put it in context. Bob Kirk calls this emphasis on thoroughly explaining the background of a breaking news story "contextual journalism."

Central TV News, with headquarters in Birmingham, describes itself as being more about people. When there was a major development in England's auto industry involving Rover, Sterling, and Jaguar, Central devoted more than 18 minutes to it on a 30-minute-long newscast. The station also effectively related the story to its viewers by complementing the talking heads with an imaginative use of animation, graphics, and other visual tools.

To cover stories in depth, Laurence Upshon encourages his Central crews to leave town, a policy similar to that of U.S. TV news. A piece on the growth of urban shopping malls meant going to the United States for angles in Pittsburgh and Oakland. For a report on cities having similar racial problems, crews were dispatched to Atlanta and Detroit.

Like many of his overseas counterparts, Upshon does not shy away from stories that are difficult to visualize. He does not tell a story by having a member of Parliament simply talk about it. Instead, Central TV illustrates the point with news videotape of what is being talked about— whether it is hospitals, schools, or roads. Upshon is also among the many foreign producers who try to tell the story from the point of view of the people affected rather than that of a bureaucrat or politician.

The BBC's Rick Thompson leans toward reporting significant news. As far as crime news is concerned, it simply "washes over you. It doesn't really affect you directly unless you happen to be one of the very small number of people affected." Thompson's example of these nonstories is a routine bank robbery in which nobody is hurt and an unknown quantity of money is taken. This is a far cry from our propensity for rolling ENG to every "barricaded man" story and hitting the air live with it.

2. *Ways to keep you awake*

Foreign producers seem to be more innovative in livening up in-depth stories by using techniques that are not only available in the United States, but probably originated here. They will use news tape, studio guests, live shots, animation footage, and imaginative graphics. TV Asahi has added miniature models to its long list of visual aids. For the coverage of a recent plane crash, the production staff constructed detailed models of the airplane and the actual terrain where the accident occurred.

Freshness is one of Kirk's prescriptions for keeping viewers awake in London. The Thames executive notes America's overuse of **file footage**, like that same ghastly scene of Challenger blowing up each time there is a follow-up or story related to that tragedy. Kirk avoids falling into the trap of using file footage for the sake of having visuals on the air. His instructions are to get out there and shoot fresh, more appropriate footage for today's story.

In the United States, we rely on shorter and shorter **sound bites.** Some are now being condensed into a dangerously incomprehensible four- or five-second timeframe. According to a study by William Edward Smith of the School of Journalism at Northeastern University, the *average* sound bite has shrunk to 10.3 seconds. The average sound bite in 1968 was 31.5 seconds. In Great Britain, France, and several other foreign countries, the newsmaker's on-camera sound statement lasts as long as it takes to make the point and/or be effective.

3. *Rediscovering that the world is round*

Foreign networks do not share our networks' fear of falling off the earth if we extend our interests overseas. As a result, they have no apparent aversion to covering news outside their international boundaries. Thus, newspeople in other countries, and their families, know a lot more about us than we do about them. It's not unusual for a seven-year-old in Örebro or a teenager in Fukishima to ask Americans about the drought in the Midwest, the ratings of the Cosby show, or the odds on the Super Bowl.

Japanese networks devote much more time to foreign news than Americans do. England's Channel 4 presents more foreign than domestic news for 50 minutes each evening. Channel 4 also consistently reports in-depth coverage of the Third World, as does the BBC and ITV.

At least two countries, England and Sweden, devote weekly scheduled time slots for documentaries produced in *other* countries, including the United States.

4. *News and ethics do mix*

Newspeople in Japan, Great Britain, and Sweden know which ethical lines not to cross. Although Prime Minister

Thatcher is accused of repressing the media, her curb on salacious and unnecessary violence has significantly cut down sensationalism on the airwaves. In addition, the BBC and ITV pursue refreshing ethical procedures without government prompting. For example, both networks in 1987 decided against showing live the righting of a capsized ferry. It was inevitable that some of the close to 200 dead bodies would be seen, and the journalists did not want the victims' relatives to go through that. Coverage didn't suffer for it. The networks professionally and efficiently reported the tragedy throughout the day with edited footage.

The Swedish Laws

In Sweden the law clearly states that one person is in charge of the news product and is answerable to any charges of libel or slander. So there's no pointing of fingers to the right or left, says national news director Odlander, who is completely and solely responsible for what is aired on the daily network "Rapport" newscast, the magazine shows, and the documentaries. Unlike in the United States, Swedish TV's modus operandi as it pertains to ethics is based on a treaty with the government that specifies objective and impartial coverage. Staging is out of the question.

Another agreement with the government makes it *mandatory* for Swedish broadcast journalists to pursue investigative reports. According to Örebro's Hermansson, "The important task for us is to look upon authorities in this society, the companies, organizations, with a critical eye." The result is reminiscent of the many fine investigative reports on U.S. channels. Örebro TV utilizes its expanded half-hour newscast on Friday nights to exercise its mission as watchdog for the people. Hermansson cites the investigative report his news department did on a scandal involving Arthur Nobel of the peace prize family. Ironically, Nobel was accused of the illegal sale of weapons to India.

5. *TV and radio can work as a team*

Sweden and England know how to combine forces for a dramatic one–two media punch. Odlander works closely with radio news in the production of his two weekly half-hour news magazine shows. A recent special dealt with Sweden's postal problems, another with AIDS. The programs start on TV and continue on radio for audience phone-in participation. The BBC, of course, has its integrated service.

In the United States, an intricate maze of network and group-owned TV and radio stations taps this obvious two-way source only in a very limited way. Among the notable examples is an experiment taking place in the Carolinas. A network of the two states' eight radio stations is simulcasting WSOC-TV's newscast out of Charlotte, North Carolina. The newsrooms will share facilities. Another exciting trend is the increasing number of public radio affiliates simulcasting the "MacNeil-Lehrer NewsHour."

6. *Anchors who can write and report*

There are certainly some serious journalist–anchors in the United States who actually contribute to their newscasts. But in some other parts of the world, anchors are always serious journalists first.

According to BBC and ITV policy, for example, smoothness in news delivery is strictly secondary to journalism know-how. As a result, few British anchors are simply pretty faces. They actually understand the stories they read, and they sound like it. And none of this has a negative impact on ratings for BBC, ITV, or France's Channel 3.

In Paris, a tired, dishevelled journalist with a suit off the rack and hair barely combed takes to the air with confidence and professionalism. Remi Guidicelli doesn't have the time or inclination to blow-dry his hair or lay on the pancake makeup. He has no problem understanding the copy or the lead-in to a reporter piece because, chances are, Guidicelli covered the story and wrote the lead-in.

Too many other countries, however, are starting to lean toward the Ken and Barbie anchor syndrome. The problem—as it was for the United States in the 1970s—is the packaging of TV news for ratings and profits.

7. Reporters with a specialty

Pleas overheard in a U.S. newsroom practically any time of the day or night: "Anybody know what *prima facie* means?" Or "What the hell is a junk bond?" Or "Quick, I need another word for angioplasty!" In American journalism, the emphasis is on the generalist, a polite term for knowing little about a lot. In European journalism, there is increasing recognition of the need to have experts in the subjects that are important to society. These include medicine, law, politics, economics, leisure, and a multitude of other topical, consumer-related areas that crop up in the news constantly.

That broadcast journalism has to respond to the needs of the people is apparent in specific, daring directions being undertaken by BBC. News editor Cramer, a leader in the movement, feels the media "unleash general reporters into an area of the world, or a story area, they know next to nothing about." Cramer, on the other hand, is more likely to hire specialists and teach them to be journalists.

In line with this policy, BBC news has set up four specialist units—bi-media units that cross radio and TV lines. The areas covered are foreign affairs, which includes defense and diplomacy; social affairs, which includes science, medicine, legal affairs, and labor relations; finances, which includes business and commerce; and politics, which includes Parliament. Experts who don't happen to have journalism training learn the ropes after being hired.

8. Saving that endangered species: Competition

The BBC's Chris Cramer also observes that TV news in the United States is becoming less competitive, while the British are still slugging it out. In the United States, for a variety of reasons, an increasing and alarming number of major news events are being pooled, meaning viewers are seeing the same coverage, regardless of which channel they are watching. At times, the process of the media electing one crew to cover a story and then **pooling** it with the others is necessary. There may be geographical, space, or security reasons for not letting in all the crews. For example, when the author produced the coverage of the pope at the Stations of the Cross in Israel, there was no way more than one crew could get into those ancient, tiny quarters without destroying them and possibly the pontiff as well. A pool was the only answer. There are several other instances, such as certain combat situations and select presidential activities, when the practice is necessary. However, pooling also cuts media coverage costs and gives the government control, factors making it tempting to invoke the emergency measure, with or without an emergency.

Rat Pack Journalism

Another undesirable practice in the United States is rat pack journalism, the tendency for the media to attend the same events, cover the same speakers, and air the same news footage and sound bites. In many U.S. cities, the local city news service budget determines what will be covered that day. Many of those items in the budget are called in to the wire service by press agents—which doesn't leave much opportunity for reportorial enterprise.

Reporters in the United States also get hooked on press club media events. At the Greater Los Angeles Press Club, for example, any person or group can stage a "news conference" for a modest fee. Too many news desks take advantage of the convenience.

In the Far East and Great Britain, coverage is more open. Cramer describes British competitors as "scrapping dogs most of the time in the field, abroad or at home." He agrees there are occasions where pooling is a necessity, "but most of the time we're at each other's throats." As a result, reporting overseas can be more challenging and the finished product more lively.

9. *Shooting down the live shot*

For certain TV operations abroad, overdoing live coverage is no problem because they don't have the high tech to do it in the first place. Sweden and Scotland, for example, still don't have their own satellite capability. But in the United States, too often a reporter will be doing a live shot in an empty auditorium hours after a speech, or standing shivering at the darkened scene of the crime well after the coroner carted off the body.

So far, the British and Japanese try to be discriminating about going live at five. The reaction from Thames's Kirk is typical. He says it's fine for a major breaking story, but when there is not much happening, "I think you can make yourself look a bit silly, actually."

An example of the Thames approach is its recent coverage of the opening of a new airport. Kirk's staffers came up with an angle involving a security problem at the just-built airport. They reported the dedication ceremony and the security problem live and complemented the coverage with a prepared in-depth background piece.

10. *Equity in the newsroom and on payday*

Most news executives abroad gasp at the amount of money we pay anchors and at the huge discrepancy between their salaries and those of other staffers. The anchor in Japan, Sweden, and England is just another member of the team. His or her job, often in addition to several others, is simply to read the news on the air. In most cases, anchors are not indispensable and many news departments have a specific policy calling for a rotation among anchors. However, as reported earlier, there are signs that anchors in other countries are beginning to emulate their U.S. counterparts and see stars in their eyes as well.

11. *A different approach in personnel*

There are some dramatic differences between foreign and U.S. work environments. In at least three areas broadcast journalists overseas do things quite differently:

1. Employees handle more than one job.

2. Producers alternate.

3. Newly hired journalists learn on the job.

Handling More Than One Job

When Japanese writer–reporters cover a story, it is assumed they know it inside out and are best qualified to edit and direct it on the air. Although it can be somewhat disconcerting to have a parade of writer–reporters step up to the console and direct their own pieces, the result contributes to smoother production. Mistakes are rare in **computer graphics**, the written information superimposed on the bottom of the screen, in the **cuing** in and out of sound bites, or in the cutting away from a newstape too soon.

In the United States, writers, reporters, and producers turn over these details to studio directors who are not normally journalists and may not have time to familiarize themselves with the content.

TBS reporter Noriko Tanaka has to be three things: journalist, reporter, and director. That means she worries her story right through air time, when she actually calls the shots and **supers**.

In the affiliates, reporters do more. At TBS's Fukishima operation, news chief Takahashi says, "You do everything." That means a reporter has to be capable of shooting, editing, writing, directing, and producing the story. While Takahashi sees the advantages in the Japanese system, he thinks it spreads his staff thin and people work "much harder than they do in the U.S."

In Italy, the reporter does it all, because of a huge vacuum in RAI's organization chart. There are no producers and no writers. That means a reporter in Italy has to take on all the responsibilities, including budgeting and ordering equipment and facilities.

Producers Alternate

At least two countries, Sweden and Japan, won't let the same person produce the show every day, which is the practice in the United States. Producer Iwamoto turns out "News Center 9," NHK's 40-minute local newscast, every other week. Iwamoto works 12 hours a day as producer, but the following week he assists his replacement, works on improving the newscast, and generally recharges his batteries.

In Sweden, the turnaround is longer. Producer Andersson is responsible for the ABC regionalized newscast every third week. During his off-time, Andersson assists the producer on duty, thinks up new ideas, and rests.

The producer burnout rate in Japan and Sweden is far less of a problem than it is in the United States.

Learning on the Job

Japan's approach to training journalists on the job instead of in school, NHK's Tanaka says, can be a nightmare. Tanaka is a journalistic live wire, but getting thrown out cold on the street with a camera crew and being critiqued by several persons afterwards "is painful."

A budding journalist applying at a TV station is told to "study, play, have fun, just enjoy your four (college) years," relates Tanaka. But she made it through the flak and today is a key reporter for TBS's local news program.

In the United States, the direction is toward increasing cooperation between academe and the profession. Many universities insist on tough requirements in the nonjournalism learning area. At the same time, there are growing programs between the media and schools of journalism. The RTNDA, for example, has become far more involved with educational institutions.   It now recruits student campus-based affiliates for a closer liaison between the two groups.

PROJECTION

When Tanaka is asked whether she would prefer working in the glamorous, big bucks U.S. TV market, she hesitates. She finds giving up a company that has hired her for life and competing in a strange environment "kind of scary."

This is in sharp contrast to the volatile United States market, where two years is considered a reasonable tenure for a reporter, anchor, or news director.

Watching reporter Tanaka work is an exhilarating experience. A bundle of energy, she plunges into her assignments with an enthusiasm that would make any U.S. news director envious. Not that it's that unusual for young reporters anywhere to show zeal as they get a crack at a challenging, exciting, and selective profession. What's different is this reporter's confidence and working environment. Tanaka has a lifetime job security that clears her head of any anxiety and, as a result, she doesn't have to spend valuable time or energy covering her back from enemies within. Tanaka, who learned the hard way, is confident when she covers a story and when she walks into a control room to direct her piece on the air. Too many of our novices are overwhelmed by the personnel, equipment, and frenzy of broadcasting live. Even seasoned producers can be intimidated by the closed shop, almost fraternity-like atmosphere of TV's warroom. However, for Tanaka, learning to direct her own stories in studio was as much a part of the training as learning how to find a story and covering it for TV.

The work environment in countries such as Japan, Great Britain, France, and Sweden is a comfortable one. Foreign newspeople seem to enjoy their jobs and each other more than Americans do. A closer equality in salaries and professional confidence are probably the main reasons for the camaraderie, but another factor is also involved. So far, the intensity overseas is funneled into doing the best possible job of gathering the news, rather than into collecting rating points. As yet, news directors in other countries don't post the overnight Nielsen and Arbitron results in the newsroom, but there's more to it than that. It's more a matter of values, of clearing the head in order to concentrate on the basic mission of informing the people.

Unfortunately, the outlook for this approach is cloudy. If the present trend continues, broadcast journalism overseas will become more like ours. It won't happen tomorrow, but at some point Asian and European newscasting will probably have the same commercial bent. As international satellite use becomes more commonplace, they will have even more opportunity to watch us and we them. That could mean a similar, delayed cycle of commercialism for the other continents.

It is also possible that by studying U.S. journalism more closely, foreign journalists will emulate the good and avoid the bad. Satellite transmission goes both ways. With more accessibility, present and future U.S. broadcasters could also be teased and tempted into taking a closer look at alternative ways of producing and airing the news.

10

---■---

THE BOTTOM LINE

The Profession in Perspective

Food for thought, an appetizer really, put together with a bit of reprise, a touch of conclusion, and a dash of suggestion . . . all of which is aimed at future journalists and designed to fortify them prior to facing the reality of looking for a job.

This odyssey's original mission was to capture for tomorrow's broadcasters a snapshot of TV news as it is today and suggest where it will be tomorrow. In doing so, it was necessary to evaluate the soothsayers' declarations of gloom and doom and chart some alternative paths for the profession. The new routes were carved out with newcomers in mind, future journalists who may be confused by the criticism and the direction of TV news.

It quickly became apparent that a lot of the sooth being said was based on a dramatically different technical and economic environment. It was a journalistic past that certainly created TV miracles, and a lot of us are very proud to have been part of that revolution in broadcasting. But a point that many executive emeriti miss is that they, too, are very much responsible for putting the industry in the precarious position it is in today.

Each time a news producer coerced management or technicians toward a live shot that hadn't existed, it brought us closer to the satellite mentality we have today. Each time a news producer harnessed creative juices to make an event exciting as well as informative and understandable, it brought us closer to the incredible popularity of news and the news-insatiable public we have today. Each time a producer squeezed that extra buck from parsimonious management, it brought us closer to the economic monster we have today.

News producers, reporters, anchors, writers, assignment editors, directors, and technicians of the 1960s and 1970s fought tooth and nail for change and they got it. The problem is that today old-timers expect to see the same shows: Edward R. Murrow doing a number on demagogue Joe McCarthy; Huntley and Brinkley relentlessly pounding out the civil rights revolution day after day; and Alexander Scourby narrating that incredibly creative history of the U.S. Navy in World War II called "Victory at Sea."

The video trail blazers also encouraged affiliates to have a more professional, more respectable news operation. For one thing, it was costly and time-consuming back in the 1960s to dispatch network producers, directors, reporters, unit managers, and technicians to breaking stories away from New York. We did it too often at too big an expense, in money as well as the health of staffers caught in a syndrome of jet lag and broken families. Many of these problems are gone today, but the solutions have sprouted migraines. However, for future journalists there are alternative routes. Some have already been pointed out—and here, they've joined with other, related thoughts.

ACCEPTING LOCAL DOMINANCE
BUT HELPING IT WORK WISELY

Those aren't UFOs whirling around up there. They are real satellites giving local stations the fantastic ability to cover the state, the country and the world. Networks have to stop being schizophrenic about it. They can't give the stations around-the-clock national and international feeds, help them buy Ku-band trucks, and then expect the affiliates to stay home.

What the networks and all of us can do is encourage local news departments to spread their wings logically. Sending local reporters to cover state delegations at national political conventions is solid journalism. Originating a local newscast there isn't. It is also questionable for a local station to send its own reporter thousands of miles away to cover the Supreme Court, summit meetings in Moscow, or a ship disaster off Puerto Rico just to show the flag. The better affiliates know it, because they have done it and grown out of it. These are stories covered competently by the networks. The Chancellors, Brokaws, Mortons, Rathers, Brinkleys, Jenningses, MacNeils, Lehrers, Moyerses, Shaws, and Williamses live and breathe politics. They've been covering political conventions, campaigns, caucuses, and stumps for generations of news. They study politics full-time prior to a political event and memorize entire space mission flight plans before launch day.

Yet the coverage horizon for future broadcasters need not be a narrowing one. Shooting out of the area, or anywhere in the world, is good journalism if it is relative to a specific audience. If a station in New York is doing a series on teenage gangs, it makes good journalistic sense to look at a city in the Midwest that has an innovative approach to the problem, another on the West Coast for the Skinheads angle, or overseas if the situation warrants.

Broadcast journalists have done it. They've hooked viewers to the point of frenzy. People watch and buy the products that are sold during the commercial breaks. As a result, news programs are assuming a status similar to that of entertainment shows—and therein lies the danger. Unless broadcasters rise above the glitz of the show biz shows, become a lot less greedy for the ratings dollar and a lot more restrained in showing off their technical abilities, their products are in danger of rising and falling just like the entertainment shows. Of course, there's nothing wrong with being commercially successful. Lincoln Steffens sold newspapers, magazines, and books. So did Ernie Pyle, Horace Greeley, and Mark Twain. But they also reported what people had to know at the time. They understood their professional and societal responsibilities.

Viewers too may soon realize that while local newscasters are preening live and in color from Korea, there may be nobody home watching the store. Viewers do get caught up in the event, whether it is the Olympics, the political conventions, or a visit by the pope. For a while they'll accept the novelty of watching coverage of a major event segue smoothly to near flawless local newscasts originating from that event. But in the newspaper the next day, they'll read about the city hall scandal, transportation crisis, or heinous crime that was going on in their very backyards at the same time and was ignored by those newscasts.

GIVING NETWORK NEWS
A DIFFERENT ROLE

Entry-level broadcast journalists do not necessarily start at local stations. Some, especially those who have had an opportunity of interning with the networks, could end up at the webs. That means they could have an active role in helping guide the wobbly network news operations toward a more positive future.

Network news departments have all the skills and techniques to fulfill a need beyond the reach of local stations. They have the expertise and the equipment to do a thorough, in-depth job of telling and analyzing significant national and international stories. They also have the talent to relate these developments, regardless of where they occur, in exciting and interesting style.

At least one commercial network is moving in that direction. ABC news president Roone Arledge in *Newsweek*, October 17, 1988, said: "We must emphasize what we do well, which is not only give the headlines but perspective on subjects not everyone is interested in." ABC's "World News Tonight" reflects Arledge's policy. Executive producer Paul Friedman is already dropping stories local viewers might have seen in favor of longer pieces.

JOINT VENTURING BETWEEN
LOCAL AND NETWORK

The time has come for the network to tap all that affiliate talent with projects benefiting the media as well as the public. No longer does the network home office have to dispatch its crews to out-of-town locations. Local broadcast journalists have improved dramatically, particularly with more seasoned network correspondents switching to the affiliate market. Networks should be taking greater advantage of this bank of talent. In addition to reporting professionally and mastering the latest in production know-how, these journalists are experts in their beats and have the confidence of their sources. The network–affiliate joint venture would create a sure-fire combination of talents and facilities that could bring a variety of meaningful stories from the grass roots to the entire country.

STOPPING THE TENDENCY TO IMITATE EACH OTHER

Network news will not solve its ratings problems by taking on the spots of Eyewitness or Action News. There is an alarmingly increasing tendency for the network anchors to slip into happy talk. The practice has become particularly noticeable on the networks' morning news shows. It was not unusual for Bryant Gumbel to ask newscaster John Palmer about his family after a serious network newscast. Now, with Deborah Norville delivering the news from the same "Today" stage set, the dialogue with Gumbel has become even more chatty. There was a time when the networks set the standards of professionalism and maturity.

Properly produced, a newscast should contain the right balance of legitimately hard and soft news without tapping the personal lives or golf games of the presenters. After all, a good newscast, like a good symphony, should have rhythm and contrast. The great composers recognized the need for a certain amount of lightness by the time the listener got to the third movement. In music it might be a change of pace, a contrasting waltz tempo or rondo. In a TV newscast, after 20-plus minutes of catastrophe, the time could be right for a light, even humorous news feature.

As for the affiliates mimicking the networks, that's an altogether too familiar strain. The viewer doesn't need a Dan Rather lookalike scampering around the world covering stories that the real Dan Rather is covering. And at least one thing CBS's $2.5-million man can't do is cover that *local* education crisis.

The "60 Minutes" ambush approach to covering a story is another technique local stations tend to imitate. Chasing down, hounding, and forcing an interview from a reluctant newsmaker is questionable, even for the CBS show. But to do it imitatively and arbitrarily risks irresponsible reporting. "60 Minutes" earned its perch among the top 10 Nielsen-rated programs for a variety of solid journalistic reasons that overshadow the ambush device.

SPENDING MORE MONEY FOR THE RIGHT CAUSES

Over the past decade or so, the spenders did get out of control and painful policing will be a necesary evil for the foreseeable future. Future broadcasters are stepping into a far less plushy environment, but like any business, unless news grows, it could die. There are causes that will have to be pushed.

Hiring More Reporters

A newsroom, typically staffed with 15 reporters, cannot be expected to cover the news as thoroughly as a neighboring newspaper with 200 reporters. If there were enough reporters, TV too could assign them to beats. It would mean giving reporters an opportunity to cover fewer stories better.

Training and Hiring Specialists

As the world becomes more complex and viewers more demanding, the need for expertise in reporting becomes urgent. Trained specialists in economics, political science, medicine, law, international relations, terrorism, crime, aging, urban planning, leisure, and areas yet to surface should be crowding out the generalists in the years ahead.

Thinking Beyond the Bird

As for the technological revolution, the momentum will continue and broadcast journalists will soon look back at satellite communication as just another landmark—like color film, videotape, or electronic news gathering. The start-up for electronic miracles to come will cost a lot of money, bucks that must be spent unless news wants to categorize itself as just another transient network sitcom or syndicated quiz show, filling in local station time.

ROBBING FROM PETER, ETC.

Without resorting to tabloid TV, there are at least two sources for funding the future of news: current anchor salaries and consultants' fees. In both cases it's going to be tough to sell management on the idea, but that's where fresh blood comes in.

News executives from coast to coast and overseas agree: On-the-air talent is being priced out of the market. That the concern is almost unanimous is a good sign. If the country's news directors and incoming leaders were completely together on this, it would be very hard for upper management to ignore the unbelievable drain on a station's budget. Those $2-million-plus salaries for network anchors and $1-million-plus for local anchors are big money. Even sports and weather anchors are nosing into the seven-figure range in larger local markets.

A more realistic salary range would deliver other perks as well. One would be a far more equitable balance among anchors and the people who are supposed to be their bosses, the news directors, and the others who put the show on the air: producers, reporters, writers, assignment editors, and technicians. The residual benefits include burying for good the prima donna complex and returning control of the newscast to the producers.

Convincing upper management, the affiliate general managers, and network presidents of all this is another matter. The strength-in-numbers routine and a profit-and-loss projection will help, as will concrete examples of the good life from home or abroad.

There's also gold in those other hills—piles of it doled out to news consultants from stations all over the United States. Not that Magid or McHugh & Hoffman have to end up homeless. A case can be made for hiring the experts for what they do best—market research. Most news operations don't have the apparatus to determine the demographic makeup of their audiences or any special, possibly overlooked needs in coverage.

But one way to trim the fat and remove a negative influence is to discourage consultants from determining newscast content and cosmetics, as well as from teaching those superficial mini-courses in journalism. That type of guidance and control leads to a sameness in newscasting. In the 1970s, consultants became enamored with the profit-heavy but journalism-weak Eyewitness News format and managed to drop its seeds at stations from coast to coast.

It's still a problem today. Consultants advise their clients to do more how-to-cope stories, which then end up proliferating through newscasts in every corner of the country. If consultants, not broadcast journalists, say weather news is out and business news is in, the viewer is left with no alternative. Consultants determine the look of the news set, the clothes anchors wear, the amount of happy talk, the emphasis on "good news," and virtually the show's entire rundown, from the opening anchor walk-on to the style of closing credits. Thanks to videotape, consultants can telecommunicate every new wrinkle to all their clients. If KING-TV's Donald Varyu came up with a new idea today on how to do sports, he insists that "two weeks from now, that tape would be on the desks of news directors throughout the country. Anything that's good is stealable. It helps feed the sameness."

Since consultants are basically market researchers and started out as such, their responsibility can be limited to that area. Of course, imaginative station executives could come up with other ways to get the marketing information. General managers could consider going directly to the source through two-way dialogue with the audience, questionnaires, public meetings, interviews, and solicitation of letters.

Still another valuable source for this talent is available to TV management. The academicians of broadcast journalism, who have been trained in research and survive on it, could be hired during break times. Many journalism educators have impressive experience in TV news as well as theoretical expertise. Their contribution can be valuable in the other aspects of consulting as well.

RECAPTURING JOURNALISTIC RESPECTABILITY

Back in the mid-1970s, news operations made a couple of near-fatal turns. One involved the responsibility of determining for viewers what it is they need to know, as opposed to letting them decide. Local newscast content and on-air talent suddenly were determined by the audience, not the news experts. Even when the subject comes up today, some news directors are likely to bristle and label the earlier philosophy "arrogant" or "pretentious." It's understandable that few executives want to admit they cater to the public's taste in order to make more money. Not as understandable are the too-few attempts to enlighten the public in palatable style on matters that are of obvious importance. Producers in the 1960s were able to do it. Producers in Europe and Asia in the 1980s and 1990s are able to do it. The problem may be what news directors *perceive* as being important to the viewer.

Sitting before live cameras for hours on end waiting for Mercury, Gemini, or Apollo astronauts to launch safely, orbit, and return to earth can be fairly dull TV. However, for producers in the early 1960s, it was standard operating procedure to prepare a variety of show elements that played during the waiting periods. We came up with Mr. Wizard–type scientists who could explain this whole new complicated world of aerospace, one simple step at a time. We came up with meteorologists, guests from the aerospace industry, and a series of tightly edited visual packages, ranging from the history of flight to a peek at aerospace in the year 2000. There were animation, graphics, live reactions from people on the street, and tabletop maps of the recovery area decorated with store-bought models of the ships standing by to scoop up the heroic pilots. It was just like FDR's World War II warroom.

In Sweden, it is important for viewers to know about their government's decision whether to join the Common Market, a boring talking head story that would hit the cutting room floor in America. In Sweden, however, broadcast journalists realize there are real people behind each aspect of the story, including workers in the affected industries and consumers in their homes, as well as corporate giants and concerned government leaders. Swedish TV produces a human story complete with people and vibrant visuals, bringing alive the country's anguish about making an historic economic decision. It also exercises a basic responsibility for informing, educating, and involving the public in an issue of great importance to it.

The other turnabout in the 1970s concerned a sudden lopsided emphasis on the show biz aspect of newscasting. Ken and Barbie were in. Gray hair was out. Unfortunately, the journalistic beauty trap survived and too many markets nationwide still perceive the country as youth-oriented.

KCOP-TV in Los Angeles took a giant step in the right direction in spring 1989 by dismissing a cosmetically appealing male anchor and replacing him with seasoned reporter–anchor Warren Olney. Fittingly, news director Ed Coghlan also changed the structure of his 10 P.M. newscast with a dramatically heavier emphasis on depth in local news.

Ironically, as management appears to get younger, the audience is getting older. The senior citizen population is growing rapidly, the logical end product of dramatically improved health and care. Its preferences in on-the-air talent and newscast content will be reflected in rating points and product buying power.

It could be a fun litmus test, using anchors and reporters who can write, report, and read news intelligently, regardless of race, creed, color, sex, looks, or age. These **newsies** might just have a fighting chance, given the same promotional and professional backup available to former beauty queens and kings.

There are other hurdles in the pursuit for broadcast journalism respectability to keep in mind:

- *The network and local news bandwagon approach to politics.* One stumble during a debate or unfavorable poll and the candidate is written off by the media. There is always a danger of the candidates' grandstanding, sensationalism, stereotyping, and fleeting images overshadowing the real issues.   Local newscasters in particular have to understand what a powerful weapon they have with their hours of news programming. Politicians realize it and take full advantage by making themselves available for free stumping time at the expense of weak interviewers and the public. Author Mark Hertsgaard has been particularly vocal in his criticism of the journalists' coverage of the Reagan administration and campaigns. He accuses them of settling into a posture of "accommodating passivity" from which they never completely arose. Instead of watchdogging the administration, the newspeople became part of the palace guard, asserts Hertsgaard. Instead of insisting on substance from the White House, they responded "with gentle, jelly-bean journalism that elevated surface over substance and obfuscated the real issues at stake." It is natural for many beginning journalists to look in awe at office holders and people in authority and be timid about challenging them. Most novices grow out of it, however, realizing how impor-tant the journalist is in protecting the public's interest.

- *The dearth of the hard-hitting, muckraking documentary and special report.* The alternative is to continue breeding generations of viewers who consider superficial "48 Hours," glitzy "Saturday Night with Connie Chung," or sensationally shallow "GER-ALDO" as in-depth, investigative journalism. "See It Now." "White Paper." "Victory at Sea." "Person to Person." They are already endangered species.

- *Untruth in advertising.* It's a given that news and mini-docs are competitive and must be promoted, but why let lies and bad taste soil the purity of broadcast journalism? Station break teases and print ads can be colorfully written, excitingly produced, and *accurate.* In a society of increasingly liberal standards, it is a challenge, but the basic truths in news can be tasteful. Hopefully, the newsroom will insist on playing a bigger role in this area as time goes by.

SWALLOWING OUR PRIDE
AND LEARNING FROM OTHERS

Radio stations in the United States outnumber TV outlets seven to one. For some reason, the two electronic media have never learned to live together despite belonging to the same family. As a result, both radio and television news miss out on an exciting potential in coverage and programming. As neither medium will ever have the resources newspapers have, consideration should be given to sharing reporters and news stories as they do overseas. **Simulcasts** have proven to be dramatically effective programming in foreign countries, particularly in the news specials category.

We should also consider:

- alternating producers and anchors as they do in Japan and Sweden;
- training producers, reporters, and writers to direct on the air;
- using imaginative tools to tell the "boring" stories;
- approaching crime coverage more maturely.

It is possible that exploitative bloodshed on TV news could be doing its part, along with superviolent movies, toward immunizing society against violence. It is estimated that by the age of 16, children have witnessed 18,000 murders on all types of TV programming. We are creating an environment in which life is meaningless and violence acceptable. By all means, the crime beat is a necessity, but it is the broadcast journalist's responsibility to cover it maturely and with perspective. Crime coverage is among the many major responsibilities future journalists will take on and undoubtedly change for the better as they make their mark in broadcasting.

However, prior to that major leap for mankind, there are some not so small steps ahead for tomorrow's newspeople, and they involve the basic challenge of getting into the business in the first place. Stay tuned.

11

NEXT STOP:
THE REAL WORLD

Charting the Mine Field
Along the Way

Help Wanted: Reporter/Photographer. Minimum one-year experience in television news . . . Tallahassee, Florida.

Anchor/Reporter for State News Network. Some experience . . . Nashville, Tennessee.

Sports Director. High energy, creative person . . . Idaho Falls, Idaho.

Photographer/Editor. Available now . . . Rockford, Illinois.

TV News Director . . . Agana, Guam.

Anchor/Reporter . . . Harrisburg, Pennsylvania.

Job seekers will find there is also TV news "help wanted" in Huntington, West Virginia; Rapid City, South Dakota; Kansas City, Missouri; Billings, Montana; and Greenville, North Carolina. The edited excerpts above are from Job Bulletin 540, the RTNDA Job Information Service, July 1989. What association members will rarely find in the monthly job bulletin board is an opening in New York, Los Angeles, Chicago, or Miami. In other words, for the fledgling broadcast journalist there is little or no action in the big city markets. Applying to those stations in most cases is a frustrating waste of time and money.

For the student, the first reality is a tough one. According to a survey done for RTNDA by Vernon A. Stone in 1989, there are 60 applications for every entry-level position in TV news. There are only some 2000 entry-level job openings a year, and metropolitan area jobs, rarely advertised, are for outstanding broadcast journalists with impressive track records. What is open in the larger cities is the flunky entry-level job such as production assistant or editorial assistant—a fancy title for copy person. Although working at a big station may impress distant relatives, it is rarely the path to a reporter or producer job. More likely, the editorial assistant gets discouraged with the low-level work and, running into a dead end, chucks it for a different field.

The challenging, beneficial, productive, and even fun part is heading to where the action does exist for the college graduate. Relatives who helped the student through college may not be too tickled with the beginning salary, however. For example, $12,000 a year for an anchor–reporter–shooter–editor job in the 200th ranked market is not unusual. But the jobs are there, and the object is to get into the process and become a working journalist. Actually, nobody expects the novice to stay longer than a year or two before jumping a peg or two up the ADI list and the salary scale. The revolving door syndrome is unfortunate in some ways, but the reality is that smaller stations can't afford to pay much more than that. They know beginners have stars in their eyes, and as they go on to the next galaxy there will always be other college graduates right behind them.

THE MARKET

As for the basic information about markets and jobs, there are several sources. The bible is *Broadcasting Yearbook*, which contains every TV and radio station, with all the vital statistics such as place in the market, address, phone number, ownership, and news director's name. TV stations are listed geographically as well as by ADI. There are a couple of catches. For one thing, the yearbooks are very expensive, but most journalism schools and libraries have them available for student use. The other problem is that news directors are perishable, so it is always safer to phone the station and double-check prior to writing.

The yearbook does not list job openings, but its magazine, *Broadcasting*, does. Of course, the entire profession is aware of it and subscribes, so it's best to get a copy and answer the ads as soon as possible. The RTNDA Job Information Service is equally popular, and it too is an excellent source for where the jobs are. Students can subscribe to the Job Bulletins for a modest fee. An even smarter move would be to join the RTNDA, thus taking advantage of all the association's periodicals and meetings. That added advantage of mixing with the professionals at meetings, conventions, and seminars pays off. News executives may not have appropriate openings at their stations, but they may be aware of opportunities elsewhere. Other sources, including computerized services, provide the job information by phone but they tend to be expensive. Details about those services are available in ads usually carried by trade periodicals such as *Broadcasting*.

There are at least two other approaches. Networking, exchanging information with alumni already working in the field, is one natural approach. The other takes a bit more initiative and imagination, but it has worked. That approach involves anticipating where there may be an opening and making contact with the stations whether or not they are advertising. In that case, there is no reason graduates shouldn't pick geographical areas they would *like* to live in. Of course, the applicant has to be realistic regarding the size of the market. It is amazing how a clever letter, phone call, road trip, or a combination of all three can open the door. It is a matter of timing and being in the right place at the right time. Indeed, results may still be discouraging, but it could be looked upon as an investment. Although there may not be a job opening today, the applicant has shown the news director an initiative and a desire to work in that particular market. And that's where the fun challenge, researching a potential market, comes in. It is an opportunity for the applicant to utilize one of the basic journalistic characteristics, curiosity. Doing that homework about the station and the community are very helpful when you approach the person doing the hiring. It also sets the mood should the applicant get hired. Learning, exploring, and mixing in a new environment can make living in a strange city a lot more enjoyable.

THE TOOLS

Specific tools are needed for pursuing employment in broadcasting. Misuse of any of the following can jeopardize what may be a once-in-a-lifetime opportunity:

- *The letter.* It has to be typed or printed out by computer, neat, concise, attention getting, free of all grammatical or spelling errors, and accurate. It's amazing how many envelopes and letters have the news director's name wrong—and end up in the wastebasket. Those applicants would have been better off if the letters had remained unread since news directors seldom forget.
- *The phone call.* Communication is the name of the game, so executives expect the applicant to speak clearly, authoritatively, confidently, and concisely. You usually get just one phone call so make it good. And for return calls, an answering machine, without rock music, barking dogs, or Ronald Reagan imitations is a must.
- *The résumé.* Neatness and conciseness are a must. Emphasize the part-time job at the TV and radio station, internship, college or professional experience, and awards. Limit it to one page maximum, keeping in mind most news directors couldn't care less about hobbies or other extraneous information.
- *The audition tape.* Another must, but not as easy to acquire. In many cases, journalism schools or internship stations will help. You can also get a tape at a commercial studio, but the process could be costly. News directors do not want an on-camera biography. They do want the *best*, quality not quantity, in field reporting and anchoring. It is far more likely the first job will be heavy in reporting—breaking news, features, investigative—so that's what they want to see. Give them black at the top and a brief black between stories. Don't go past 15 minutes, clean up all the mistakes, and include a typed list of the pieces, including length and cues.

Much more can be said and written about audition tapes, résumés, and the other tools necessary for the job pursuit. Each person preparing for the plunge is unique, with different experience, talent, specialties, interests, and goals. Also unique is each geographic and market area, with different demands, policies, and relationships with the university. In other words, the best follow-up to these basic prerequisites is to pick the brains of the experts within reach: teachers, advisers, and working broadcast journalists.

WHAT THE INDUSTRY WANTS

The professionals have been vocal in what they demand of incoming employees. News directors surveyed by the RTNDA were recently asked to list the factors most important to them in evaluating job applicants. They are, in descending order:

1. Appearance
2. Writing skills
3. Previous work
4. Hands-on experience

A journalism-communications degree is ranked seventh, followed by "a well-rounded liberal arts background."

A different survey, listing the things news directors look for in applicants, ranks knowledge of current events first, followed by proper enunciation, good listening skills, and ability to deal with news sources. In terms of most important general course areas, English is first, followed by government, political science, economics, history, and logic. Radio–television–film is seventh.

THE REAL BASIC: WRITING

Writing is the weakest skill in the newsroom. You hear this same complaint in New York, Boston, Minneapolis, Salt Lake City, Seattle, Portland, and San Francisco. News directors want to take writing ability for granted, but they can't. They, and the industry as a whole, are finally starting to realize that without good writing there is no good television news.

In the beginning, writing was the basic force that generated broadcast journalism, as it did print journalism. But over the years, its pedestal has slowly eroded away. Ironically, success—the explosion in TV news and high tech—was a major factor in its downgrading. Feeding the public's hunger for news meant shifting priorities. High tech, not writing, became the great god—and quantity and speed became the priorities.

But the two, style and speed, can live together very compatibly *if* the student broadcaster is made aware that writing, not white balance, is the core of any form of journalism. Although **white balancing,** getting the camera ready to record colors accurately, is essential to shooting a story, there is no story without writing. These days it is easy to get carried away by the toys, but high tech should only be considered a means to an end.

The basic role of the journalist is to tell a story, a *complete* story with a beginning, a middle, and an end. That ability, storytelling, is what generates the machinery. With that ability, writers learn how to research, outline, and unravel information coherently, concisely, and conversationally:

- Research. Research doesn't mean news writers spend days in the library poring over books and going through term paper mechanics before turning in their copy. But it does mean that writers take advantage of any information available to them prior to putting a story together. That could include easily accessible wire service copy, newspaper files, a quick phone call, and that important fund of information already stored in the mind from constant reading and seeking. The research phase is an absolute must prior to interviewing anyone for a story. It helps the interviewer ask intelligent questions and gauge the veracity of the answers.

- Outline. The outline is simply a device for organizing one's thoughts—a checklist, something to change and manipulate until the order makes sense. For big projects, like documentaries, the outline must be thorough and detailed. For a story near deadline, it may be only three or four scribbled notes. For the breaking story at the scene, the outline should become automatic after a certain amount of experience.

- Completeness. For a professional journalism product, writers write, reporters write, and, in another sense, producers write and editors write. The principles remain the same. The story put together in the newsroom or at the scene has to have the beginning, middle, end. The producer must keep those same three elements in mind in designing the newscast. Editors cannot edit videotape without them.

Since broadcast news is a direct medium, the writing has to be conversational and not encumbered by long sentences or dangling phrases. Unlike the newspaper, which can be read and reread, this message is aimed at the ear as well as the eye. Those who master it are those who work at it constantly—in class, out of class, in the internship, out of the internship, before graduation, after graduation. They don't simply type their stories or rewrite material line by line. Instead, they get into the habit, from the beginning, of *talking* their stories into the typewriter or computer. They also learn how to determine the point of the story and relate that as if they were talking to a friend.

Some applicants tend to ease up on spelling, grammar, and neatness once they've left college. But the rules are just as important for broadcasting as they are for print, and errors are red flags to news directors. Basic mistakes in these may indicate sloppiness in other vital areas such as in gathering facts and meeting deadlines. Without accomplishing simple, mistake-free news writing it is unlikely the cub will reach the next plateau, which is the development of a distinctive style.

Robert Mulholland has very specific gripes about contemporary broadcast writing, particularly about the technique of writing "voice-over"—the anchor or reporter narrative copy accompanying the visual scenes. He is one of those concerned with how technology has negatively affected writing style. Reporters and writers now write their stories and have editors put in covering visuals afterward. Like so many writers from the 1960s, Mulholland was taught to first edit the piece visually, making sure the scenes flow and everything works. Now "some guy (the videotape editor) drops in scenes" and that means the story may not make sense. Visuals are often inserted only because they are available and needed to cover the copy. In an effort to have the story move faster on the air, editors also tend to cover the copy with a multitude of short scenes that don't necessarily relate. That results in a repetition of shots used and overuse of file film. Mulholland cites Persian Gulf coverage and the overuse of the same shot of "the four Iranians with the one machine gun in the boat because it's the only shot we have of the Iran suicide squad."

In the golden age of writing, a story was told with a combination of pictures, words, and sounds. But today, the tendency is toward a sort of show and tell, an assault of visuals for the sake of visuals.

Future broadcast journalists should learn both methods of writing for videotape. They should certainly learn the contemporary way of writing the story first, then adding the visuals. But it could add to the quality of the product if they were also capable of putting together the video prior to writing the script. That technique is still used in most documentary writing.

As for students seeking instruction and guidance in how to write well, there are steps that can be taken. They should determine first whether the journalism program is accredited. The qualifications of the teachers should also be checked to determine if they had experience writing and producing in TV news. This type of information is normally available in journalism department offices and catalogues. The better schools have relationships with the profession in the form of internships, seminars, and guest lecturing. Most pros in those environments are approachable. At the right time and the right place, they would be willing to review copy with the learning writer.

Good TV writing, like good TV news, rarely comes naturally and has to be learned. It is carefully researched, cleverly constructed, conversational, concise, and colorful, with the givens of neatness, accuracy, and orderliness. And the payoffs are exciting. For one thing, the standards are the same, whether the end product is a print story, newscast, documentary, or movie. In other words, it opens the door to the whole world of writing. There is also a close relationship between the art of professional broadcast news writing and the overall product. Writing could be a catalyst in improving the quality and standards of TV news.

GETTING A JUMP ON THE REAL WORLD

Students should be on the lookout for schools that stress the real-world environment and are staffed with instructors who know the business. Hopefully, the profession and the schools will continue efforts to better interrelate. The creation of RTNDA affiliates on campuses is a good move, as is the program developed by the National Association of Broadcasters in conjunction with the Broadcast Education Association. Through that joint venture, instructors have the opportunity to work as summer relief in TV newsrooms during break periods and sabbaticals.

As for the student, the internship at TV and radio stations is invaluable exposure to the real world—but the quality of that process is crucial. Some newsrooms provide a learning experience for students; others take advantage of interns by having them do menial jobs. Even minimum wage is a rarity, and too often there is no financial commitment. A good internship provides an opportunity to learn the work environment, get professional guidance, make contacts, earn money, and pepper the résumé with real call letters.

Many schools have their own radio and TV stations and students should take advantage of these as well. Writing, reporting, editing, producing, and anchoring real morning and afternoon drive-time news for a real radio station is invaluable experience. It too looks great on the résumé. At the very least it hammers home punctuality and deadlines, which in the real world are a matter of employment life or death. If the university doesn't have a license, a station wired up for the campus will do. In many instances, students have taken the initiative and become instrumental in creating campus stations. If that is not feasible, some low budget commercial radio stations would welcome the opportunity to get free news programming.

There is also a multitude of opportunities for students in TV. Universities everywhere are taking advantage of the multichannel structure of cable and are already producing news product on a weekly or daily basis. Low-power TV is still another potential outlet for student product. FCC-approved permits for broadcasting over a limited geographical area, with a maximum radius of 25 miles, are far easier and cheaper to get than commercial licenses. Yet, commercials are permitted on LPTV, so the station can also generate income.

Writing, reporting, editing, producing, and anchoring real newscasts and documentaries under real deadline and professional pressures accomplishes several other purposes. In addition to bolstering the résumé, it makes the product eligible for awards competition and creates an outlet for coverage benefiting both town and gown.

Another ongoing demand of the real world is knowing what is going on all the time. Reading newspapers and magazines, listening to radio news, and watching TV news should be a continuous, self-assigned homework assignment whether school is in or out.

Overkill? News directors consistently bemoan the newcomers' illiteracy in current events, lack of knowledge about the field of broadcast journalism, limited information in noncommunications subjects, and naive attitude about broadcasting work habits.

LEARNING THE COMMITMENT

Donald Varyu says: "They don't know what it is to have to work for a better shift or get a story when somebody doesn't want to talk to you."

James L. Boyer says: "The kids I see coming out of college . . . frankly don't give a damn. We bring in college interns and the first thing they want to know is when is lunch hour."

Real newsies simply don't work 9 to 5, and when there's work to be done, an 80-hour week is not unusual. Many of them started out by putting a radio station on the air in the morning, signing it off at night, and doing whatever they could in between. And they did it just to learn everything there is to know as fast as possible, just for the opportunity to get a foot in the door of a remarkable profession.

It could have something to do with an overall society enraptured with materialism. Or the "me" syndrome. Or the well-publicized romanticizing of a handful of anchor stars and their multimillion-dollar contracts. Or it could be the lop-sided influence of profitmaking on the part of the people who own the broadcast companies. Young people, as we learned in the strife of the 1960s and 1970s, have a knack of recognizing a problem and doing something about it. They could make a broader, more meaningful approach to their chosen field of broadcast journalism, beyond the rush of fame and money.

After all, *they* are tomorrow's news directors, producers, writers, reporters, and assignment editors. Their influence will be felt positively and historically if they just take a good look at a heritage that goes back to Thomas Paine and that spawned contemporary saints from Murrow to Moyers. They can also take a critical look at what's right and wrong about their chosen field. In addition, they can take

- a more questioning, challenging look at government and all forms of authority;
- a more compassionate look at the societal problems that surround us;
- a more imaginative look at ways to inform, educate, interpret, and even entertain;
- a more realistic look at delivering the payload without forcing the employer into bankruptcy.

Former President Jimmy Carter has noted that questions asked at presidential news conferences have expanded from 15 words in Franklin D. Roosevelt's term to 150 today. His reaction is a clue to what ails broadcast journalism today:

One of the things you must remember is that the questioner is often more interested in the question than the answer.

The next generation of broadcast journalists could start by learning, and change the profession by doing. It is a profession that should, again, be dedicated to cynicism but not narcissism, to iconoclasm and not only commercialism.

Having dedicated themselves to brilliant writing; accurate, objective reporting; fearless muckraking; and devotion of time and body to the cause, future broadcast journalism supermen and superwomen need just one other requisite. Humor. Living out of a suitcase, working around the clock, eating off wax paper, missing birthdays and anniversaries, and dodging bullets can be a bit discouraging without it. The ones who make it manage somehow to enjoy the craziness and the rewards.

And that can be as simple as the pride of turning a neat phrase, or the exhilaration of rewarding experiences such as a clean beat over the opposition, a polished product after an impossible set of circumstances, the glory of being in the award winners' circle, or the look on the face of a person whose life you have changed for the better.

GLOSSARY

Action news Newscast format emphasizing quantity of items, a fast pace, and action videotape such as crime news.

ADI Area of dominant influence; a method of evaluating a station's market area on the basis of geography, demography, and viewer habits.

Affiliate Independently owned TV station contracted to carry the network's programming.

Arbitron Arbitron Ratings Company, one of the two major researchers determining the quantity and demography of viewers for ratings.

Assignment editor Newsroom person responsible for coordinating and assigning reporters and camera crews to cover the news.

Bean counter Slang for executives responsible for cutting costs or keeping them down.

Beat A journalist's or newsroom's designated coverage area, such as a police or fire department, city hall, or a neighborhood, or a specific subject, such as crime, economics, sports, or medicine. Or, an exclusive story or piece of information that other journalists don't have. In the good old days it was called a "scoop."

Beeb Britons' affectionate nickname for the publicly owned BBC, the British Broadcasting Corporation.

Bell and Howell Portable film camera used for covering news events.

Bird Slang for satellite.

Book See **Ratings book**.

Breaker News story that is developing, or breaking, at the present time.

Cast Another word for newscast.

Computer graphics Computerized method for enhancing the presentation of a TV news story by providing visual backup, including diagrams, titles, and animation, for display in the window next to the anchor person, superimposed on the picture or on the full screen by itself.

Cuing Signaling an action, such as starting or stopping a videotape, an on-camera speech, a movement, or the operation of a camera. The word *cue* can also refer to story script information, such as videotape length and sound content, essential for the studio director and smooth flow of newscast.

Dissolve One picture gradually appearing on the screen, replacing one that just as gracefully disappears.

Downlink Signals transmitted from a satellite down to a ground receiver. Or, the dish that receives the signals.

Editing bay Cubicle-sized workroom in which videotape editor edits incoming news tape from the scene into size, shape, and order suitable for on-the-air.

ENG Electronic news gathering; creates the ability to transmit live or taped signals over distance instantaneously.

Feeds Electronic material transmitted to the studio or newsroom from other sources, such as a news scene, news service, or network.

Fiber optics Newly developed technique for transmitting huge amounts of video and audio information economically and efficiently by earth ground cable.

File footage Newsroom's library of videotaped or filmed news stories aired on prior newscasts.

Graphics All-purpose term referring to any number of visuals—such as maps, charts, illustrations, or animation—used to illustrate a news story.

Happy talk Self-serving, "ad-libbed," inane banter among anchors and in-studio reporters during a newscast.

Hard news Timely, significant news events, as opposed to soft, human-interest stories.

HDTV High-definition television; Japanese-developed system that doubles the number of lines making up a video picture, creating the sharpest, clearest TV pictures yet.

Instant news special The process, started by NBC News in the 1960s, of putting together and airing a news show on a major development as fast as possible.

Junk news New wave of pseudo-news shows produced outside of news departments by independent production companies and usually featuring sensational, highly commercial subjects.

Ku-band Section of the electromagnetic spectrum assigned to satellite use and thus used to transmit news material to and from the satellite.

Ku-band truck Satellite news vehicles from which stations can shoot pictures to satellites, which shoot them back down again.

Lash-up Technically bringing together several different elements for a common purpose. For example, a lash-up could include broadcast lines from New York, Washington, and Pittsburgh for an on-the-air conference or interview.

Lead First, normally most important, story in the newscast.

Lineup Outline, rundown, routine, or order of stories and production elements in a newscast.

Live Broadcasting an action as it is happening.

Local-local Covering news closer to the station's home base rather than farther away.

Lower-third supers Written information, including names, titles, dates, and locations, displayed at the bottom of the screen during a newscast. Also see **Supers**.

Magazine show TV news show format featuring longer, hopefully more carefully prepared stories.

Market Geographical area made up of the potential audience served by the TV station.

Microwave As opposed to cable, high-frequency radio signals through which electronic information is transmitted.

Mini-docs Serialized version of a TV documentary spread out over several newscasts, often in order to rope in viewers during competition for ratings.

Mixer Device that can bring in several sources of audio or video materials and create one output.

Modem Accessory permitting sending and receiving of materials, such as research, scripts, and lineups, over great distances by phone line from computer to computer.

Muckraking In *Pilgrim's Progress* (1678), it meant, literally, raking up the muck. In President Teddy Roosevelt's time (1906), it meant hard, digging, investigative reporting in search of political, government, or business corruption.

Narrowcasting Gearing program content and advertising toward a specific racial, economic, social, religious, or other special interest group.

Network A group of interconnected stations, some of which may be owned by the network, that broadcasts the same programs—for example, ABC, CBS, NBC, PBS, CNN.

News budget Breakdown of expenditures involving specific or long-range news projects. Or, list of stories compiled by wire services or newsroom.

News director Executive in charge of news department product and personnel.

News junkie Person who puts watching news above food and family, refers to anchors by their first names, and takes CNN intravenously.

Newsie Professional, trained journalist who is serious about the job and the responsibility.

A.C. Nielsen One of the two major research companies determining the quantity and demography of viewers for ratings.

O & O TV station actually owned and operated by the network.

Paint boxes Lightning fast, computerized system of producing graphics for visual support of news stories.

Pooling Method by which competing media agree on having one reporter and crew cover story and sharing results.

Preempt To replace a TV program with another one, as in the case of a major news special canceling an entertainment show.

Producer Person in charge of creating, coordinating, and putting on a newscast.

Promos Commercials of various lengths designed to promote station's or network's own news product or personalities.

Rat pack journalism The result when competing reporters and camera crews cover the same story from the same angle with the same results.

Ratings book The book in which the rank order, determined by research companies such as Nielsen and Arbitron, of viewers' preferences of specific competing programs is given. Results, based on sweeps, determine price to be charged per commercial minute.

Reverse question Technique, which can be considered staging, of reshooting the reporter's questions from an interview with or without the subject being present.

Robotic camera An automated way of televising newscasts that is being used more and more, and in which no hands, or people, are involved.

Show rundown See **Lineup**.

Simulcast A program presented simultaneously on TV and radio.

Skin test A test to determine chemical and emotional reactions to TV news personalities or programs by attaching electrodes to viewers and registering results on graphs.

SNG Satellite news gathering; the instantaneous method of covering and transmitting a news event over distances and obstacles by bouncing it off the bird.

SNV Satellite news vehicle. See **Ku-band truck**.

Soft news Light, featury, frothy human interest stuff heavy in cute pets and children.

Sound bite Taped and edited on-camera statements by newsmakers.

Sputnik First successful man-made satellite launched by the Russians in 1957.

Stacking Order in which the stories appear in the newscast.

Staging Reenacting an event for the camera, usually without admitting it is a fake.

Standuppers Videotaped or live reporters delivering part or all of their stories on-camera at the scene of the story.

Story count The number of stories in a newscast.

Supers Visual written material that is superimposed over already existing live or videotaped picture. As a result both images are seen simultaneously. Also see **lower-third supers**.

Sweeps Month-long, four-times-a-year competitive madness during which two major ratings services survey viewers to determine most popular programs. See also **Ratings book**.

Syndication Service composed of news, sports, weather, and feature stories provided by the networks directly to the stations for use in their local newscasts.

Tabloid news See **Junk news**.

Talking heads Sight and sound of newsmakers, often shown in head-and-shoulders shots, talking on camera without benefit of accompanying illustrative footage.

Tease A hint about the story coming up with just enough about it so viewers suffer through the commercial and come back for the rest of the show.

TelePrompTer Device that projects the script in front of the camera lens, making it look like newscasters memorized the newscast.

Teletext Method of transmitting text materials and graphics to the home TV screen.

Telop Outdated technique of projecting still pictures into newscasts.

Trash TV See **Junk news.**

Uplink Signals transmitted to a satellite for relay elsewhere.

Videotex Method of receiving text materials, graphics, other visuals, and services on home screens through interaction between viewer and source.

Viewfinder Eyepiece, or lens, into which camera operator peers to set up the focus and composition of a shot.

VNR Video news release; complete, ready-to-go-on-the-air "news" stories produced and distributed by news sources.

VTR Videotape recording.

Webs Slang reference for networks.

White balance Method of adjusting camera to the light for best possible color value.

Wire services Companies that cover news stories of all types throughout the world and sell them for distribution to TV, radio, and newspaper clients. These include general services such as Associated Press, strictly foreign ones, and those specializing in subject areas such as sports or weather.

INDEX

157